I0010253

Game Development Patterns with Unreal Engine 5

Build maintainable and scalable systems
with C++ and Blueprint

Stuart Butler

Tom Oliver

BIRMINGHAM—MUMBAI

Game Development Patterns with Unreal Engine 5

Group Product Manager: Rohit Rajkumar

Publishing Product Manager: Vaideeshwari Muralikrishnan

Book Project Manager: Sonam Pandey

Content Development Editor: Debolina Acharyya

Technical Editor: Simran Ali

Copy Editor: Safis Editing

Proofreader: Safis Editing

Indexer: Tejal Daruwale Soni

Production Designer: Gokul Raj S.T

DevRel Marketing Coordinators: Nivedita Pandey, Namita Velgekar, and Anamika Singh

First published: December 2023

Production reference: 1071223

Published by
Packt Publishing Ltd.
Grosvenor House
11 St Paul's Square
Birmingham
B3 1RB, UK

ISBN 978-1-80324-325-2

www.packtpub.com

To my wife, Becky, for her unrivaled love, support, and understanding as I pursue my career goals, and for always being by my side throughout all of the challenges. To my boys, Jack, Zach, and Lincoln, for their support and excitement at the prospect of me writing an actual book.

– Stuart Butler

To my wife, Rosie – her unconditional love and support for my efforts to reach my goals continue to keep me sane and true. To my parents, Martin and Sheila, for giving me the foundation I needed to succeed. To my sister, Amy, and her husband, Phil, for inspiring me to drive for better.

– Tom Oliver

Foreword

Stuart Butler and *Tom Oliver* are two of the most talented game developers and educators that I have had the pleasure of working with. Both are experts in the technologies that underpin modern video games and have a wealth of experience in turning imagination into reality through the creative sorcery of design and code.

Like most coders, my first programming project was getting the terminal to print out "Hello World." From here, I started to tinker, making increasingly more complex and challenging projects in a very solution-focused manner. My code was often a sprawling mess of comments and redundancy; it wasn't very efficient, but it worked (usually). It wasn't until I had to revisit (and decipher) old code or if I had to collaborate with fellow developers that my cluttered code became a problem. Fortunately, my teacher introduced me to the world of design patterns.

Design patterns are one of the most valuable assets in the game developer's arsenal. They keep your workflow manageable and efficient, and common patterns make collaboration easier. In the fast-paced, multi-disciplinary world of game development, design patterns are an essential ally.

Through this book, you will learn the essential principles of programming and build a strong understanding of the value and application of design patterns. You will benefit from an appreciation of how these approaches work in practice through working with one of the most powerful tools in game development, the Unreal Engine 5 games engine. You will learn the essential grounding to allow you to seamlessly blend between the Unreal Blueprint visual scripting system and C++, allowing you to develop complex and scalable video game projects.

Butler and Oliver are exceptional developers, designers, and world-leading experts in the Unreal Engine. The knowledge they share through this book will be invaluable to anyone working with this tool. Their step-by-step approach and the strong focus on practical examples and learning through doing will help you get the most out of your game's development projects.

Unlock the full potential of the Unreal Engine with Games Development Patterns.

Prof. Christopher J. Headleand

National Teaching Fellow, Head of Games Development,

Staffordshire University

Contributors

About the authors

Stuart Butler is an Unreal Engine expert with over 13 years of experience in teaching game development in higher education. Stuart has published projects in a multitude of disciplines, including technical design, art, and animation. Stuart is the course director for games technology at Staffordshire University, responsible for the programming team within the UK's largest games education department. Stuart is also an Unreal Authorized Instructor and has published work as an educational content creator with Epic Games, developing learning materials for Unreal Engine 5. Stuart holds a BSc (hons) in computer games design and a PgC in higher and professional education.

I would like to thank my family for their unwavering support and my colleagues at Staffs Uni for inspiring me to pursue my ambitions. I'd also like to thank Greg Penninck, Bobbie Fletcher, and Justin Mohlman especially, for believing in me throughout my career so far and providing some amazing opportunities. Lastly, I'd like to thank Tom for joining me on this adventure to becoming published authors.

Tom Oliver is a game programmer with over 10 years of experience in working with game engines, both commercially and in an educational capacity. He has used Unreal Engine for contract work both in and out of the games industry, from creating systems for games to mixed reality training simulations. Tom is a senior lecturer and course leader at Staffordshire University, responsible for maintaining the award-winning structure and teaching of the course in the UK's largest games education department. Tom holds a BSc (hons) in computer games design and programming and a PgC in higher and professional education. Tom specializes in researching gameplay systems driven by mathematical phenomena.

I would like to thank my family for always being my tribe and my colleagues at Staffs Uni for being my squadron. Special thanks to Yvan Cartwright and Paul Roberts for starting the programming fire in me – I hope that this book shows that I paid attention, and Davin Ward, for sticking with me for the last decade. Lastly, thank you to Stu for navigating these word-filled waters with me – we did the thing.

About the reviewer

Ahmed Farrag, a proficient game developer with an MSc in computer science, excels in Unreal Engine development and problem-solving. He is currently the Unreal Engine lead developer at TransformologyXR, building on a rich background that includes contributions to Rumbling Games Studios, Astra Nova, Epoch, and Camouflage Studios.

Ahmed's portfolio boasts a range of titles, including the successfully published titles *Vulcan Runner* and *Knights of Lights: The Prologue*, and the upcoming releases *Atlantis Heroes* and *Kingdom Sports*. His unwavering commitment to innovation in delivering exceptional interactive experiences makes him a standout in the industry.

Table of Contents

3

UE5 Patterns in Action – Double Buffer, Flyweight, and Spatial Partitioning

4

Premade Patterns in UE5 – Component, Update Method, and Behavior Tree

Part 2: Anonymous Modular Design

5

Forgetting Tick

6

7

Part 3: Building on Top of Unreal

8

9

Structuring Code with Behavioral Patterns – Template, Subclass Sandbox, and Type Object 177

10

Optimization through Patterns 205

Preface

Welcome to *Game Development Patterns with Unreal Engine 5*. In this book, we will be exploring design patterns, a series of tools and practices through which we can learn to write faster and easier to work with code. We will also be exploring a range of different patterns and learn to apply them to project development in Unreal Engine 5.

By the end of this book, you will be able to design systems with the perfect C++/Blueprint blend for maintainable and scalable systems.

Who this book is for

This book is targeted at beginner and intermediate game developers who are working with Unreal Engine and would like to improve their C++ coding practices. This book will help you produce clean, reusable code using design patterns. We will be covering some introductory tasks to show the key fundamentals of using Unreal Engine 5 and some of its tools; however, we will not be teaching you Unreal Engine from scratch.

You would benefit from having some experience with Unreal Engine 4 or 5, but you do not need a deep working understanding of the toolset.

What this book covers

Chapter 1, Understanding Unreal Engine 5 and Its Layers, explores Unreal Engine 5 and offers a brief history. We will look at the "fuzzy" layer that bridges the gap between C++ and Blueprint and learn how to translate Blueprint back to C++.

Chapter 2, "Hello Patterns", focuses on the principles that underpin all good code. We will explore design patterns as well as some common Blueprint mistakes, looking at how we can fix them.

Chapter 3, UE5 Patterns in Action – Double Buffer, Flyweight, and Spatial Partitioning, discovers how Unreal Engine 5 employs these three design patterns as we explore a range of tools within the engine.

Chapter 4, Premade Patterns in UE5 – Component, Behavior Tree, and Update Method, utilizes the pre-built implementations of these three design patterns and explores the tools within the engine to expand simple systems.

Chapter 5, Forgetting Tick, develops your understanding of Tick, looks at why its usage can cause issues, and explores two approaches to building systems without it.

Chapter 6, Clean Communication – Interface and Event Observer Patterns, explores design patterns that allow us to improve how different actors communicate with each other, producing more efficient solutions to communication.

Chapter 7, A Perfectly Decoupled System, discovers how we can use UML as a methodology for planning class hierarchies, to decouple the reference train.

Chapter 8, Building Design Patterns – Singleton, Command, and State, examines these three design patterns to understand their usage, limitations, and suitability across a range of game genres.

Chapter 9, Structuring Code with Behavioral Patterns – Template, Subclass Sandbox, and Type Object, explores the three most common structural patterns while building weapons classes in C++, which we will expand with Blueprint, exploring how the two languages can be used together.

Chapter 10, Optimization through Patterns, dives into the key elements of optimization before releasing games by exploring the Dirty Flag, Data Locality, and Object Pooling design patterns.

To get the most out of this book

You will need a version of Unreal Engine 5 installed on your computer. All code examples have been tested on Unreal Engine 5.0.3, and they should work with later versions of the Engine. However, this may not be the case if Epic Games makes any major changes to the core engine.

Software/hardware covered in the book	Operating system requirements
Unreal Engine 5	Windows
Visual Studio or JetBrains Rider	Windows

If you are using the digital version of this book, we advise you to type the code yourself or access the code from the book's GitHub repository (a link is available in the next section). Doing so will help you avoid any potential errors related to the copying and pasting of code.

We've included commented versions of the code found within the book as part of the GitHub repository, as opposed to including comments in the code samples, making the code easier to read and follow within the book.

Download the example code files

You can download the example code files for this book from GitHub at `https://github.com/PacktPublishing/Game-Development-Patterns-with-Unreal-Engine-5`. If there's an update to the code, it will be updated in the GitHub repository.

We also have other code bundles from our rich catalog of books and videos available at `https://github.com/PacktPublishing/`. Check them out!

Conventions used

There are a number of text conventions used throughout this book.

`Code in text`: Indicates code words in text, database table names, folder names, filenames, file extensions, pathnames, dummy URLs, user input, and Twitter handles. Here is an example: "This also applies to the `ScoreWidget` class we will use to display the player's score, which has been provided as part of the `Chapter Resources` folder."

A block of code is set as follows:

```
class APlayerController_CH7 : public APlayerController
{
public:
    void Init();

protected:
    UPROPERTY(EditAnywhere)
    TSubclassOf<APawn> _PlayerPawn;
    UPROPERTY(VisibleAnywhere, BlueprintReadOnly)
    TObjectPtr<ACharacter_CH7> _Character;
}
```

Bold: Indicates a new term, an important word, or words that you see on screen. For instance, words in menus or dialog boxes appear in **bold**. Here is an example: "Enable the checkbox next to **Editor symbols for debugging** and click **Apply**."

> **Tips or important notes**
> Appear like this.

Get in touch

Feedback from our readers is always welcome.

General feedback: If you have questions about any aspect of this book, email us at customercare@ packtpub.com and mention the book title in the subject of your message.

Errata: Although we have taken every care to ensure the accuracy of our content, mistakes do happen. If you have found a mistake in this book, we would be grateful if you would report this to us. Please visit www.packtpub.com/support/errata and fill in the form.

Piracy: If you come across any illegal copies of our works in any form on the internet, we would be grateful if you would provide us with the location address or website name. Please contact us at copyright@packtpub.com with a link to the material.

If you are interested in becoming an author: If there is a topic that you have expertise in and you are interested in either writing or contributing to a book, please visit authors.packtpub.com.

Share Your Thoughts

Once you've read *Game Development Patterns with Unreal Engine 5*, we'd love to hear your thoughts! Please click here to go straight to the Amazon review page for this book and share your feedback.

Your review is important to us and the tech community and will help us make sure we're delivering excellent quality content.

Download a free PDF copy of this book

Thanks for purchasing this book!

Do you like to read on the go but are unable to carry your print books everywhere? Is your eBook purchase not compatible with the device of your choice?

Don't worry, now with every Packt book you get a DRM-free PDF version of that book at no cost.

Read anywhere, any place, on any device. Search, copy, and paste code from your favorite technical books directly into your application.

The perks don't stop there, you can get exclusive access to discounts, newsletters, and great free content in your inbox daily

Follow these simple steps to get the benefits:

1. Scan the QR code or visit the link below

https://packt.link/free-ebook/9781803243252

2. Submit your proof of purchase
3. That's it! We'll send your free PDF and other benefits to your email directly

Part 1: Learning from Unreal Engine 5

In this part, we will be exploring Unreal Engine and the design patterns that are already included in the engine, or available to use as part of the many tools within the engine.

We will start by exploring Unreal Engine 5 and how it works with layers of code, before exploring a series of common mistakes that developers make when using Blueprint and seeing how to fix them. We will then discover a range of different patterns, exploring these with examples built from a mix of Blueprints and C++ and using a variety of tools, including World Partition and AI Behavior Trees.

This part has the following chapters:

- *Chapter 1, Understanding Unreal Engine 5 and Its Layers*
- *Chapter 2, "Hello Patterns"*
- *Chapter 3, UE5 Patterns in Action – Double Buffer, Flyweight, and Spatial Partitioning*
- *Chapter 4, Premade Patterns in UE5 – Component, Behavior Tree, and Update Method*

1

Understanding Unreal Engine 5 and its Layers

Design patterns are a series of tools and practices by which we can learn to write faster and easier-to-work-with code.

Applying design patterns to projects developed in **Unreal Engine 5** (**UE5**) will allow you to make your projects more performant, easier to read, and build upon, as well as develop an improved understanding of how the engine works.

We're going to begin by developing an understanding of the history of Unreal Engine and ensuring that we are all set up to work with the engine, covering some basic ideas of how C++ and Unreal Engine's visual scripting language, **Blueprint**, are linked.

In this chapter, we're going to cover the following main topics:

- Introducing Unreal Engine 5
- Installing Unreal Engine 5 and preparing your development environment
- The "Fuzzy" layer – bridging the gap from C++ to Blueprint
- Translating back from Blueprint to C++

Technical requirements

Before embarking on this journey of discovery, know that this book will assume some working knowledge of C++ syntax and the Unreal Engine Editor. Familiarity with pointers and how to follow code in your chosen **integrated development environment** (**IDE**) will be key to understanding the Unreal core API.

You will require the following software:

- Unreal Engine 5 (this book has been written with version 5.0.3).
- **Visual Studio** is a decent free IDE (basic support for the engine is present, meaning projects may show errors and *IntelliSense* may not auto-complete some keywords, but the project will compile and run).
- If you have access, JetBrains Rider version 2022 or later has built-in support for Unreal Engine, which will make the development process a lot easier. Rider is an alternative IDE to Visual Studio that is often preferred among programmers working with Unreal. It offers improved support for working with C++ in Unreal Engine, including auto-complete. You can learn more about it by visiting `https://www.jetbrains.com/lp/rider-unreal/`.

Introducing Unreal Engine 5

Unreal Engine 5, or Unreal for short, is a game engine developed by *Epic Games*. Unreal, as with any other game engine at its core, simply processes data from files and instructions into data that you can see on the screen. The suite of tools provided are designed to assist you in creative and predictable tasks. Having been created for you by elite programmers and designers, it is like outsourcing your development to the big leagues.

However, like a high-performance racing car driven by a learner, even the best tools can perform badly. The experts at Epic didn't know how you would use their tools. So, when we design our code architecture, we need to keep this break in communications in mind. This is where this book comes in, teaching the expected best practices when writing games in Unreal Engine 5.

Unreal Engine has powered an impressive list of gaming titles covering a vast array of genres, shipping on a multitude of platforms.

Unreal Engine 3 powered some of the biggest hits from the seventh generation of game consoles, ranging from third-person shooter games such as the *Gears of War* series developed by *Epic Games* themselves to fighting games such as *Injustice: Gods Among Us* and *Mortal Kombat* from *NetherRealm*, as well as strategy games including the *XCOM* series developed by *Firaxis Games*.

The eighth generation of consoles saw Unreal Engine 4 expand its portfolio to include racing titles such as the *MotoGP* games from *Milestone* and *Assetto Corsa Competizione* from *Kunos Simulazioni*, as well as facilitating the hugely popular introduction of the Battle Royale genre with *PlayerUnknowns' Battlegrounds* (*PUBG Studios*), which is listed as the fifth highest-selling video game on *Wikipedia*, and *Epic Games' Fortnite*, which transitioned to Unreal Engine 5 in December 2021 as detailed in *Chapter 3*.

Unreal Engine 5 features a series of new, key technologies, including *Nanite*, a sub-mesh level of detail system allowing massive polycount models to render with even screen-space hull size at any distance, and *Lumen*, a real-time lighting solution that mixes mesh distance fields and atlassed local surface data progressively over time to create a realistic lighting effect at low cost. These new technologies drove the transition from 4.26/4.27 to 5.0 as part of a complete rework of Unreal's rendering technology. The engine had an impactful debut with the release of *The Matrix Awakens: An Unreal Engine Experience*, a demo featuring photorealistic likenesses of Carrie-Anne Moss and Keanu Reeves, and an open-world city featuring impressive examples of crowd and traffic simulations.

Increasingly, games studios are opting to replace proprietary in-house developed game engines with Unreal Engine 5; studios such as *CD Projekt Red* (*The Witcher*) and *Crystal Dynamics* (*Tomb Raider*) have announced the decision to use Unreal Engine 5 in the latest installments of their game series.

Now that we've covered a little about Unreal Engine's past, pedigree, and influence on gaming history, we will get things set up next so that you are ready to use Unreal Engine 5 with C++.

Installing Unreal Engine 5 and preparing your development environment

Unreal Engine can be downloaded either via the Epic Games Launcher (available from `https://unrealengine.com`, which will install the engine for you) or as source code from GitHub (`www.github.com/EpicGames`), allowing users to compile the engine and modify it to fit their projects.

To engage with the activities in this book, you won't need to compile the engine from source unless you really want to. The benefits of compiling from source will likely come much later into your journey of working with Unreal Engine and C++. You will, however, need to install version 5.0.3 (or above) of the engine and have an IDE installed. This section covers the download, installation, and setup of the engine from scratch using the Unreal Launcher and installation of Visual Studio 2022 Community. If you already have the engine and Visual Studio installed, you can skip over this section.

Firstly, you will need to download the Launcher from `https://unrealengine.com` by clicking the **Download** button in the top-right corner of the page and clicking the **Download Launcher** button on the following page.

Once downloaded, you will need to install the Launcher from the `.msi` installer.

The Launcher is home to the Epic Games Store, your library of purchased and downloaded games as well as versions of Unreal Engine (4 and above). You will need to navigate to **Unreal Engine** and then **Library**, then click the *yellow plus icon* and select your engine version from the resulting engine slot:

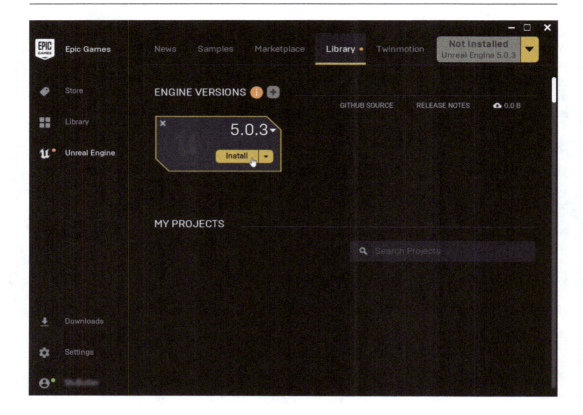

Figure 1.1 – The default Epic Games Launcher with no versions of the engine installed

Alternatively, if you don't have any versions of the engine installed, you can click the *yellow drop-down button* in the top right of the Launcher application.

Both approaches will present you with the option of where you would like to install the engine. It is advisable to install the engine on an SSD if possible as the engine will load significantly quicker than from an HDD:

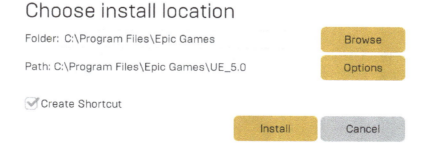

Figure 1.2 – Install location options

Clicking the **Options** button will present you with a series of optional elements for installing the engine:

Unreal Engine 5.0.3 Installation Options   ×

Core Components (Required)	32.84 GB	
Starter Content	664.93 MB	✓
Templates and Feature Packs	3.27 GB	✓
Engine Source	251.63 MB	✓
Editor symbols for debugging	50.40 GB	

Figure 1.3 – Installation options for Unreal Engine 5.0.3

We advise installing **Starter Content**, **Templates and Feature Packs**, and **Engine Source**.

Engine Source will make browsing and debugging code easier but does not allow you to rebuild the engine; for that functionality, you will need to download the source from GitHub, as mentioned earlier.

Editor symbols for debugging allows the debugging of C++ in the Editor. This is not required but will prove useful to facilitate jumping from the Editor to engine code and allow you to explore the code behind Blueprint nodes.

If you decide you want to add/remove elements later, you can modify these choices by clicking the *down arrow* next to **Launch** on the engine slot and selecting **Options**:

Figure 1.4 – Location of the Options menu on an engine slot

Enable the checkbox next to **Editor symbols for debugging** and click **Apply**.

Once you have the engine installed, you can move on to installing Visual Studio.

You will need to download the Visual Studio installer from `https://visualstudio.microsoft.com/downloads/`.

When you run the installer, select the **Game Development with C++** preset in the **Workloads** tab and select optional components, as shown in *Figure 1.5*:

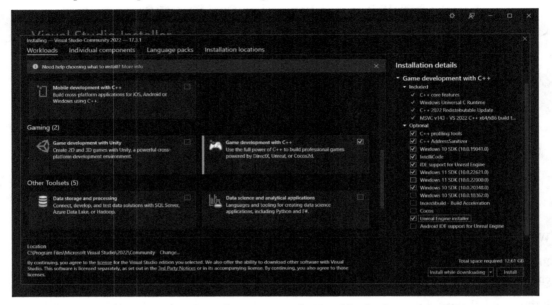

Figure 1.5 – Suggested installation options for Visual Studio Community 2022

Once you have Visual Studio installed, you should be ready to go. If you want to further improve your experience of working in Visual Studio with Unreal Engine 5.0.X, you can find some additional guidance from *Epic* at `https://docs.unrealengine.com/5.0/en-US/setting-up-visual-studio-development-environment-for-cplusplus-projects-in-unreal-engine/`.

Now that we've installed the engine and set up our IDE, we can start exploring the link between C++ code and Blueprint.

The "Fuzzy" layer – bridging the gap from C++ to Blueprint

Most engines work on the concept of layers. Each layer has a specific job to do, and when stacked in the correct order, they simplify the development and maintenance of an engine. For example, if a new graphics language emerges that it would be wise to support, only the layer with the graphics pipeline needs to be changed.

There is no hard and fast rule for the number or breakdown of layers in an engine, but what you will generally find is a separation that resembles *Table 1.1*:

Tool
Full of Editor functions that speed up standard tasks. This suite is generally what an engine is marketed on.
Gameplay
All of the custom systems created to facilitate interaction mechanics that will be bespoke in the built game.
Function
Where all automatic internal systems are handled like input capture and physics processing, etc.
Resource
Memory management and asset streaming are handled here.
Platform
Essentially, the graphics pipeline definition and build platform native integrations if the engine supports it.
Base
Full of core dependency libraries such as UI frameworks for the Editor and math libraries.

Table 1.1 – Common engine layers and their uses

Although Unreal may not explicitly label its layers in this way, we can see them in action in the relationship between how Unreal processes C++ and Blueprint gameplay. Functions created with certain specifiers in C++ can be accessed in Blueprint, but the reverse is not true. This shows there is an order to our actions where signals can only be passed one way. We'll refer to this internal separation between gameplay layers as the *Fuzzy layer*.

The fact that the Fuzzy layer exists places a limitation on how we design our systems, but in turn, we gain a separation that enables gameplay programmers to work alongside designers with little friction. Systems can be developed for simple creative use within accessible Blueprints with the more efficient C++ code hidden out of sight. To facilitate this construction, Unreal gives us **Property Specifiers** and **Function Specifiers** to define how signals will punch through.

Property Specifiers

Property Specifiers define characteristics of C++-defined variables when viewed and accessed in the Blueprint layer. The Unreal docs provide a handy table explaining the different levels of security afforded by each, along with some more specific ones designed for events, collections, and replication over networks (`https://docs.unrealengine.com/5.0/en-US/unreal-engine-uproperties/`). The six Display Property Specifiers most commonly used are as follows:

- `EditAnywhere` – The value will be changeable in all class defaults and instance detail panels. This specifier is generally used while prototyping as it displays the variable in most places, with the most options for changing its value. However, security is the price paid for this flexibility, allowing any designer to change the default and instance values, and so access should be restricted down to what you actually need once a system is tested as working.

- `EditDefaultsOnly` – The value will only be changeable in class defaults. Useful when a variable needs to be changed for balancing and all instances are spawned at runtime, where you wouldn't have access to instance detail panels anyway. Can also be used to ensure no spawned instance has a rogue different value if necessary for execution.

- `EditInstanceOnly` – The value will only be changeable in instance detail panels. Useful for allowing designers different values on bespoke placed actors in a scene but restricting the default value to something that is tested as working.

- `VisibleAnywhere` – The value will be visible in all class defaults and instance detail panels, with no option for changing it from the Editor. This is useful for debugging how the initialization process affects a value when it is unknown if the code is generally wrong or wrong at an edge case. The latter will show incorrect values at the instance level, whereas the former will be wrong at both levels.

- `VisibleInstanceOnly` – The value will only be visible in instance detail panels. Useful for surface-level debugging of values in each instance without cluttering the screen with debug messages when you have a large number of instances spawned.

- `VisibleDefaultsOnly` – The value will only be visible in class defaults. Useful for designers to reference what a functional value is and create a parity in the visual elements of an actor. This is the highest security level as each actor will display the starting value in one place.

There are two access specifiers we need to pay attention to for now: `BlueprintReadOnly` and `BlueprintReadWrite`. They give child Blueprint-based classes access to either just the getter or both getter and setter in their graphs.

Function Specifiers

Function Specifiers work similarly to Property Specifiers, defining how functions should be seen and accessed by the Blueprint layer, with some subtleties to their usage. You can find a full list of Function Specifiers in the Unreal docs (`https://docs.unrealengine.com/5.0/en-US/ufunctions-in-unreal-engine/`), but the three we are interested in are detailed next:

- `BlueprintCallable` – As the name suggests, Blueprint classes can call this function if it is in a parent class and it has the correct encapsulation type (public or protected).

- `BlueprintImplementableEvent` – The stub for this function signature is defined in C++ without any implementation. This allows C++ systems to call it and Blueprint systems to fill out its actual body. You might use this for triggering visual effects like a laser beam on a gun when it is fired.

- `BlueprintNativeEvent` – This allows C++ to define a function that is filled out in Blueprint, but in this case, there can also be a *default* implementation, which will also be run. Unreal achieves this by generating two more function definitions for you: `*_Implementation()` and `Execute_*()`. The former is used for the C++ side that must be run, and the latter is the function that must be called in C++ to fire both implementations.

> **Important note**
>
> As with layering, the C++ side of `BlueprintNativeEvents` will execute before the Blueprint side.

Using Property and Function Specifiers, we can make systems that cross the Fuzzy layer, but almost as important as routing function signals is designing inheritance hierarchies that smooth this process.

Useful inheritance

As standard practice, it is best to make sure that anything instanced in your world is a Blueprint class. This helps with debugging and linking classes with references as you have all the visual tools built into the Editor for tracing executions, and if classes are renamed or if they move directories, then links are live instead of text-based, preventing crashes.

To make this inheritance strategy work, it is recommended you think about your system from an abstract gameplay point of view. Which classes affect the mechanics? These classes need to have a C++ representation for efficiency, and so a hierarchy can be designed. From there, we inherit Blueprint classes from the end of each branch. This gives us the Blueprint classes to link, create instances of, and add visual components to:

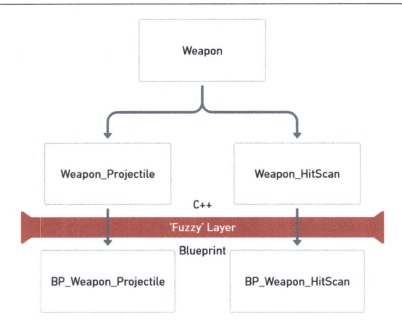

Figure 1.6 – Example hierarchy for a weapon mechanic that includes
both Projectile and HitScan mechanic types

In *Figure 1.6*, the C++ classes would contain all the logic for ammo, damaging actors, interaction handling, and so on. Then, in the Blueprint classes, components to display the mesh, run the kickback animations, and display the muzzle flash would be added. There would most likely be a function in the `Weapon` parent class called from somewhere in the firing logic that looks like this:

```
UFUNCTION(BlueprintImplementableEvent)
void Fire_Visuals();
```

This could then be implemented in the Blueprint classes to trigger visual effects. Using some of the patterns we will cover later, namely the type object pattern, you can create a vast array of weapons from these few classes with some simple asset and variable changes. This shows how the artist, designer, and programmer can all work together on one system without getting in each other's way while still benefitting from the efficiency gain in C++.

In theory, this process is perfect, but theory rarely translates to the real world. A lot of dev time is usually spent working out how to do something based on forum posts, documentation, and videos. It's great that these resources exist; almost every problem you come across has likely been solved by someone else, but there is no guarantee they have been developed with the same practice. You will come across a situation where the fix you need is in a Blueprint tutorial/example and your system needs it to be in C++, so let's have a look at the translation process.

Translating back from Blueprint to C++

Equipping yourself with tools to translate back from Blueprint to C++ is the smart thing to do, not only for the everyday fixes as mentioned previously, but also for the triple-A setting in a large company. A typical Unreal mechanics development pipeline at a large studio might follow the following process:

1. A designer has an idea for a mechanic and is given time to prototype it in Blueprint as a **proof of concept**.
2. This mechanic is tested to see if it should be developed further.
3. If it receives a green light, it will be a programmer's job to convert the prototype into a more robust C++ system.

We've covered the tools for designing C++ systems to work with Blueprint effectively, but when faced with the aforementioned situation, how do you take another person's vision and translate it into something that works efficiently yet doesn't lose what made it good? Well, the obvious answer is to directly translate it. Node for node. That process is easy, as nodes in Blueprint are quite literally just C++ functions that you have to find in the engine source. Let's take a closer look:

1. The first step is to hover the mouse over the node. Every node has a `Target` class, and mousing over will tell you what that is.
2. Next, we go to the Unreal Engine docs (`https://www.unrealengine.com/en-US/bing-search?`) and search for *U< Target class name>::<Node name with no spaces>*.
3. The Unreal docs page for the function is likely to be bare, but it will give you the `#include` directory for the file header containing that function. Once a class has been included, it can be used and explored via autocompleting a dot accessor.

> **Important note**
> If the target is `Actor`, then it will have an A instead of a U at the beginning, as shown in the example in *Figure 1.7*. This is one of the oddities of UE5; each time there is one, it will be mentioned in a box like this.

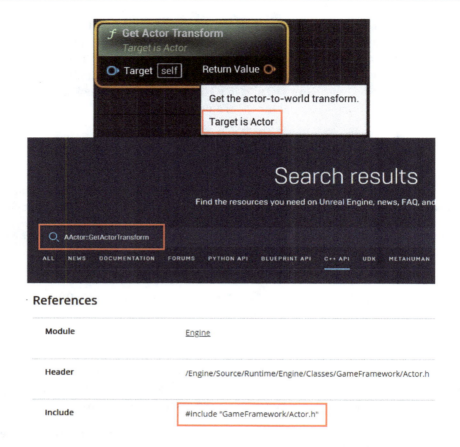

Figure 1.7 – Showing the process from node to Unreal C++ documentation

This process may seem tedious but after a while of searching different things, you will start to see patterns and be able to work out where functions are likely to be. Each node with an execution pin (the white arrow) then becomes one line of code with input pins forming the function arguments, and output pins are generally the return type (multiple of which would form a struct).

> **Tip**
>
> Reading through the following headers will be useful for mastering the translation process as they are the most frequently used: `AActor.h`, `UGameplayStatics.h`, and `UkismetSystemLibrary.h`.

The next question is, how do you know where to leave the line between C++ and Blueprints? Theoretically, you could make everything in C++, but as we've already shown, that is not a good idea for teamworking reasons. The answer has already been hinted at, but generally, you want everything visual or not to do with the gameplay mechanics to be Blueprint-based. Where that line exactly sits is up to interpretation and not worth agonizing over.

Worked example

To cement this theory, let's work through an example. In *Figure 1.8*, you can see an example Blueprint system designed to increment an integer when it is overlapped by a projectile; then, once it reaches 10 overlaps, it will play a sound and unhide a particle effect:

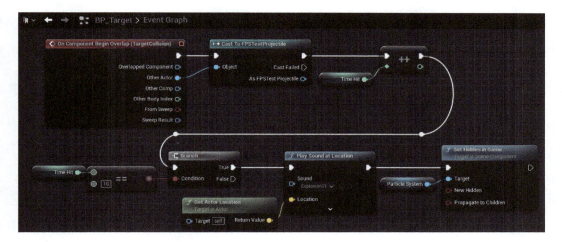

Figure 1.8 – A series of Blueprint nodes in a basic function; this piece of Blueprint has been arranged using a reroute node for readability but would normally be laid out more linearly

First, we need a C++ class for this to inherit from, and seeing as it is called `BP_Target`, we can name it `Target`. The `Target` class will need some variables; we can tell from the event that there is some kind of collision component named `TargetCollision`, which has its overlap event bound to this function. As a component, it needs to be stored in a pointer; otherwise, we would be referencing the class, not an instance. We also need an integer named `TimesHit`. As discussed prior, mechanics are made by a team of programmers, designers, and artists. Linking the right particle system for user feedback is a designer's job, so we'll leave that as Blueprint for now, but we do need a way to fire the Blueprint side, so a `BlueprintImplementableEvent` will be needed. With that in mind, the header for our new class will look something like this:

Target.h

```
#pragma once

#include "CoreMinimal.h"
#include "GameFramework/Actor.h"
```

```cpp
#include "Target.generated.h"

class USphereComponent;

UCLASS(Abstract)
class EXAMPLE_API ATarget : public AActor
{
    GENERATED_BODY()

    UPROPERTY(EditDefaultsOnly)
    USphereComponent* _TargetCollision;

    int _TimesHit;

public:
    ATarget();

    UFUNCTION()
    void OnTargetCollisionBeginOverlap(AActor* OtherActor,
        int32 OtherBodyIndex, bool bFromSweep, const
        FHitResult& SweepResult);

    UFUNCTION(BlueprintImplementableEvent)
    Void TenHitVisuals();
};
```

> **Important note**
>
> Notice how the UCLASS() call includes the Abstract tag. This will stop designers from accidentally creating instances of your C++ classes that have not had any visuals set up.

The next step is to populate the function bodies:

Target.cpp

```cpp
#include "Target.h"
#include "Components/SphereComponent.h"
#include "ExampleProjectile.h"

ATarget::ATarget()
{
    _TargetCollision =CreateDefaultSubobject <USphereComponent>
        (TEXT("TargetCollision"));
```

```
    _TargetCollision->SetupAttachment(RootComponent);
    _TargetCollision->
        OnComponentBeginOverlap.AddDynamic( this,
        &ATarget::OnTargetCollisionBeginOverlap);

    _TimesHit = 0;

}

void ATarget::OnTargetCollisionBeginOverlap
    (UPrimitiveComponent* OverlappedComponent, AActor*
        OtherActor, UPrimitiveComponent* OtherComp, int32
        OtherBodyIndex, bool bFromSweep, const FHitResult&
        SweepResult)
{
    ExampleProjectile* otherProj =
        Cast<AExampleProjectile>(OtherActor);
    if (otherProj != nullptr)
    {
        _TimesHit++;
        if (_TimesHit == 10)
        {
            TenHitVisuals();
        }
    }
}
```

The constructor is going to be standard, creating the default sub-object for the collision component and binding the overlap function to that component's overlap event (OnTargetCollisionBeginOverlap). In the overlap function, the cast node becomes a cast to a temporary "cache" variable with an if statement to check its value against nullptr. _TimesHit can then post increment, and the branch converts to an if statement which, if passed, will call the BlueprintImplementableEvent to pass the signal to the Blueprint child class.

> **Note**
> You are not required to build this example; it is here for reference only.

There will be plenty more examples of Blueprint to C++ conversion throughout the rest of this book, but if you would like to see some first-party examples, the template projects created by Epic can be loaded in both C++ and Blueprint versions.

Summary

In this chapter, we have discussed what Unreal Engine 5 is and how to get it set up with a useful development environment. We also defined some of the key terms and tools we will be using in later chapters and shared a best practice strategy for building mechanics.

Lastly, we showed the process of translating from other practices to our new best practice.

This forms the foundation for designing well-structured systems. At this point, you should be able to at least plan mechanics for your game and implement something that utilizes both C++ and Blueprint.

In the next chapter, we will cover the process of optimizing a bad project by implementing some of the patterns we will see later in the book. This should give you an understanding of why design patterns are useful before we jump into a deep dive of each one.

Let's end the chapter by answering a few questions to test our knowledge and cement some of these practices covered in the chapter.

Questions

1. Which one of these describes the implementation of `BlueprintNativeEvent`?

    A. A function that can be described in Blueprint.

    B. A function defined in C++ with both C++ and Blueprint implementations.

    C. An event declared in C++ but used in Blueprint with an `Execute` function.

2. How does Nanite work?

    A. Magic

    B. Mesh distance formulas

    C. Sub-mesh **LOD** groups

3. Which one of these is not a Property Specifier?

   A. `BlueprintWriteOnly`

   B. `EditInstanceOnly`

   C. `VisibleAnywhere`

4. Which base class adds an A to the start of each of its children in C++?

   A. `GameplayStatics`

   B. `Actor`

   C. `ActorComponents`

5. All C++ functions should be marked as which of the following?

   A. `BuildOnly`

   B. `Pure`

   C. `Abstract`

Answers

1. B
2. C
3. A
4. B
5. C

2
"Hello Patterns"

This chapter will focus on the principles that underpin all good code. Design patterns are an extension of these principles, so it is important to walk before you run. Once the basics have been covered, we'll look at some bad Blueprint code so that we can understand the process of fixing it, which we will do by following some step-by-step guides.

In this chapter, we'll be covering the following topics:

- S.O.L.I.D. principles
- Exploring solutions to common problems
- The trade-off

Technical requirements

For this chapter, you will need a blank UE5 project open and ready. There's no need for C++ right now; we'll be focusing on Blueprint code for simplicity.

The files for this project can be found in the `chapter2` branch on GitHub at `https://github.com/PacktPublishing/Game-Development-Patterns-with-Unreal-Engine-5/tree/main/Chapter02`

If you've not created a blank UE5 project before, the following steps will take you through creating a simple Blueprint project, which is all we need for this chapter:

1. Click the **Launch** button in the top right of the **Epic Launcher | Unreal Engine | Library** tab, where we installed the engine in the first chapter.
2. Select **Games** on the left side of the **Unreal Project Browser**.
3. Select **Blank** from the main section, and toggle **Starter Content** off. This will prevent Unreal from adding a bunch of unrequired assets to the project.

4.   Choose a sensible location (the default is fine) and set the project name as `HelloPatterns`:

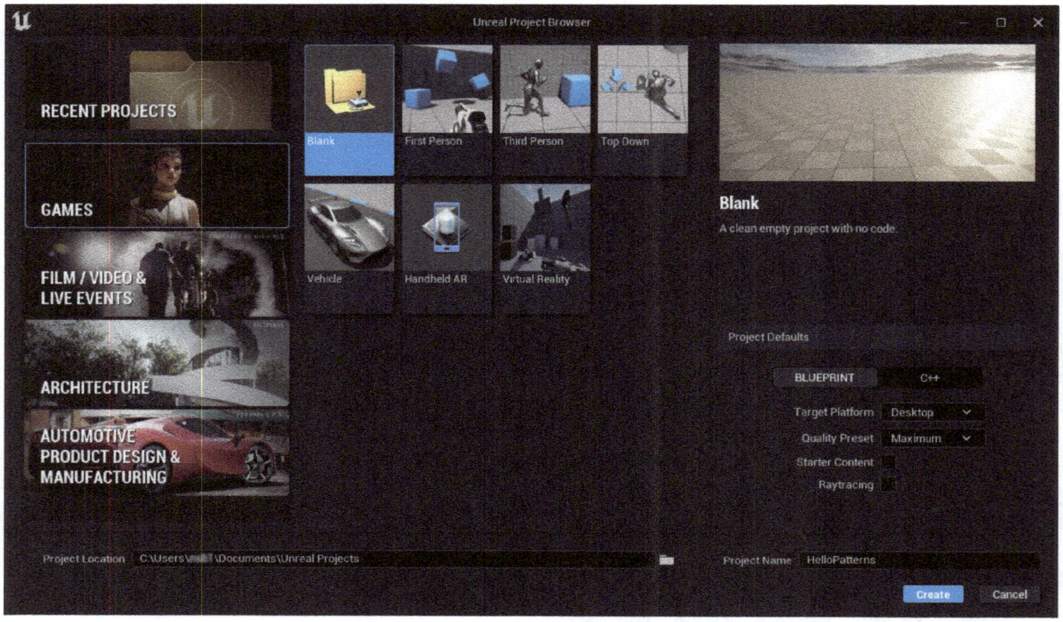

Figure 2.1 – The Unreal Project Browser with a new, blank games project set up for creation

5.   Click **Create**.

Now you have your project set up, you are ready for the interactive part of this chapter, but first, let's discuss what we mean when we say *good code*.

S.O.L.I.D. principles

To understand why design patterns work, we need to understand the principles that underpin good code. Let's run a whistle-stop tour through most people's first experience with code.

The universal *Hello World* program and the beginner exercises that follow are all made up of linear code. Linear code executes each line sequentially through a single file in which the whole program is written. Loops, selections, and statements make up literally everything.

As each program gets larger, the code becomes unsustainable. The solution is object orientation. **Object-oriented programming (OOP)** adds a lot, primarily classes and objects, as the name would suggest. With OOP's added complexity, it becomes quite easy to accidentally build in fundamental issues with systems that make it impossible to expand them. Having a set of rules to follow greatly simplifies the process of building code that works for you and the next person to maintain it. These rules take the form of the **S.O.L.I.D.** principles, which are defined here:

- S – **Single responsibility**
- O – **Open-closed**
- L – **Liskov substitution**
- I – **Interface segregation**
- D – **Dependency inversion**

Single responsibility

In *Agile Software Development: Principles, Patterns, and Practices*, Robert C. Martin said the following:

> *A class should have one and only one reason to change, meaning that a class should have only one job.*

Each class should *do what it says on the tin*—or, in this case, the class name. If you were hiring a plumber, you wouldn't expect that plumber to also recite *Hamlet* while fixing the pipes. The same is true for code; if you have a math library class, you expect it to take numbers in and return the results after performing some equations. What you don't expect is for it to use these numbers as a seed to generate an image, which it sets as your desktop background. Granted, this is an extreme case that may even be considered a virus, but the point stands.

Writing a class that does what is expected by its name helps when working in a team. Games can become quite large with many interconnected systems, and if a developer can understand what a class is responsible for and how to operate it, from just function and class names, then time is saved. Implementation can be as simple as consistent naming conventions combined with compartmentalized functionality. For instance, a function called `AddItem(ItemType type, int amount)` on an inventory component in the object used for our player probably adds an amount of an item to the inventory of the player. You don't need to open the function to see what it does due to this clear naming.

Open-closed

In *Agile Software Development: Principles, patterns, and practices*, Robert C. Martin said:

> *Objects or entities should be open for extension but closed for modification.*

New functionality should be easy to create, without needing to modify the existing code. It is easy to see this in action through something like a save system. If you want to port your game to multiple platforms, then it would make sense to have a different `save` method for each. To start with, there are two target platforms (PC and Xbox); the save system might look like this:

Example SaveClass.cpp

```cpp
void SaveClass::SaveGame(GameProgress* gameProgress)
{
    if(target == "PC")
    {
        //save game progress the PC way
    }
    else if (target == "XBOX")
    {
        //save game progress the Xbox way
    }
}
```

Somewhere down the line, you decide to extend the system to also work on PlayStation. This is another platform to handle. The modification seems easy—just add a new case to the ever-expanding statement. Therein lies the problem; the more platforms we support, the longer this will get, and the more time and memory the function will occupy:

Example SaveClass.cpp

```cpp
void SaveClass::SaveGame(GameProgress* gameProgress)
{
    if (target == "PC")
    {
        // save game progress the PC way
    }
    else if (target == "XBOX")
    {
        // save game progress the Xbox way
    }
    else if (target == "another platform")
    {
        // save game progress on another platform
    }
    else if (target == "yet another platform")
    {
        // save game progress on yet another platform
    }
}
```

The solution is to spin the `save` function into an abstract function with an overriding child class function for each platform shown in *Figure 2.2*. This not only means new platforms can be added with a new child class but also, we don't have to instantiate versions of the save system we will never use, saving memory:

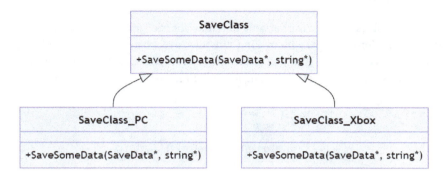

Figure 2.2 – UML diagram showing inheritance of a SaveClass
instance with the open-closed principle applied

Next, we will look at Liskov substitution.

Liskov substitution

In *Agile Software Development: Principles, Patterns, and Practices*, Robert C. Martin said the following:

> *Let q(x) be a property provable about objects of x of type T. Then q(y) should be provable for objects y of type S where S is a subtype of T.*

Essentially, if there is a child class in use within your code, you should be able to swap it out for any of its siblings without modifying the structure of the calling code or using variables for unintended purposes. In the previous example, we turned a save system into a parent-child structure. If we were to add cloud saving, we would no longer have a file path to send data to and instead, we would use an IP address. Well, we could make this work by just piping the IP address through the file path argument as shown in the following exerpt but this is a violation of our naming convention. Undoubtedly, someone down the line will need to use this function and won't understand what they need to do. Plus, any data not in the requested type will have to be converted by the calling object, sent, and then parsed inside the child implementation, which wastes time:

Excerpt from a program calling a save system

```
SaveClass* _XboxSaver = new SaveClass_Xbox();
SaveClass* _CloudSaver = new SaveClass_Cloud();
SaveData* _DataToSave;
```

```
//Some code to prepare save data

_XboxSaver->SaveSomeData(_DataToSave, "filepath");
_CloudSaver->SaveSomeData(_DataToSave, "127.0.0.1");
```

The fix for this principle is to make the variable that may change a member of the child class. That way, initialization functions can deal with making sure the correct data is asked for, and we can easily swap out save objects as needed. This also means the abstract functionality from the parent can be called the same way wherever it is used, making for easier maintenance as systems grow:

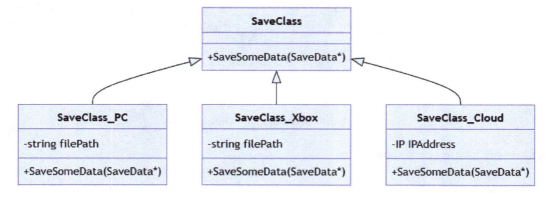

Figure 2.3 – UML diagram showing inheritance of SaveClass with Liskov substitution principle applied

So, if we apply the Liskov substitution to our code:

Excerpt from a program calling the better save system

```
SaveClass* _XboxSaver = new SaveClass_Xbox("filePath");
SaveClass* _CloudSaver = new SaveClass_Cloud(new IP());
SaveData* _DataToSave;

//Some code to prepare save data

_XboxSaver->SaveSomeData(_DataToSave);
_CloudSaver->SaveSomeData(_DataToSave);
```

As you can see, the calls become the same format and are therefore substitutable and satisfy the principle.

Interface segregation

In *Agile Software Development: Principles, Patterns, and Practices*, Robert C. Martin said the following:

> *A client should never be forced to implement an interface that it doesn't use, or*
> *clients shouldn't be forced to depend on methods they do not use.*

If the parent defines an abstract function, the child must override it. This causes issues when implementing an abstract function for some of the child classes as the other children that don't need that functionality must override to immediately nullify it. The solution can either be to extend the inheritance so that there is another layer for the classes in need of this extra functionality or to create an interface (only available in multi-inheritance languages) that adds the functionality to certain subclasses.

Going back to the save system, if you need to get account authorization before saving on Xbox and PlayStation, then this can be done via inheritance. A new sub-class of `Save` can be made, which declares an abstract function for getting authorization. This then becomes the parent for each save system that needs the function:

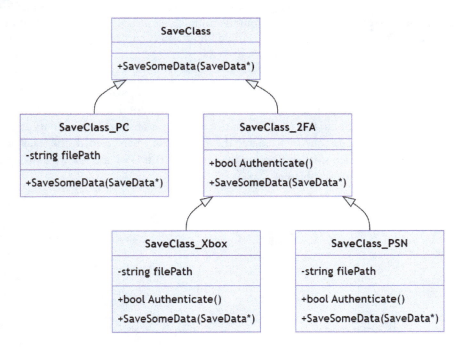

Figure 2.4 – UML diagram showing SaveClass inheritance with
interface segregation applied through pure inheritance

So, if we apply the concept of interface segregation to our code:

Excerpt from a program that checks authentication before saving

```
SaveClass2FA* _PSNSaver = new SaveClass_PSN("filePath");
SaveData* _DataToSave;

// Some code to prepare save data

if (_PSNSaver->Authenticate())
{
    _PSNSaver->SaveSomeData(_DataToSave);
}
```

> **Important note**
>
> Technically, the structure shown in *Figure 2.4* will violate the previous Liskov substitution principle as sub-classes that don't inherit from the middle *interface* layer can't be substituted into the pattern. Deciding which principle is more important is part of the skill of designing code and is usually learned with practice.

Although this inheritance structure does satisfy the **interface segregation** principle, it does end up getting quite messy with lots of segregated behaviors. There is another way. Applying the first principle of single responsibility to this tree, we remove each segregated behavior into its own class. Objects of this can then be created and injected into the save system where needed. The result is more independent classes, but cleaner, smaller inheritance structures. This is what is known as using **class composition** to solve a structure problem:

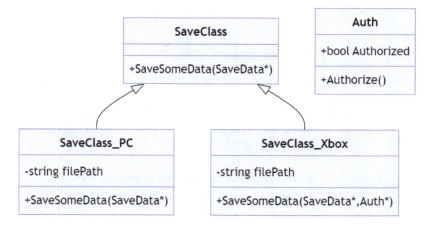

Figure 2.5 – UML diagram showing SaveClass alongside Auth hierarchy

So, if we apply class composition to our code:

Excerpt of a program that implements Figure 2.5

```
Auth* _Authorizer = new Auth();
SaveClass* _XboxSaver = new SaveClass_Xbox("filePath");
SaveData* _DataToSave;

//Some code to prepare save data

_Authorizer->Authorize();
_XboxSaver->SaveSomeData(_DataToSave, _Authorizer);
```

The `SaveSomeData` function would then check the authorized status of the `Auth` type object as it needs to.

Dependency inversion

In *Agile Software Development: Principles, Patterns, and Practices*, Robert C. Martin said the following:

> *Entities must depend on abstractions, not on concretions. It states that the high-level module must not depend on the low-level modules, but they should depend on abstractions.*

This pattern is the most useful for UE5. Simply put, if you inject a reference to a class somewhere, it really should be the highest-level abstract parent that still has the functionality you need. Easy examples would be in a PlayerController, where you don't reference a specific pawn sub-class but, instead, the idea of a pawn. Better than that, you could reference the interface for the functionality you need; this will prevent long cast chains to filter for the type.

In the save example, the injection of the account authorizer is a concretion that assumes every `save` module will authorize in the same way. In reality, it would be much more appropriate to inject an `Authorizer` parent that only contains the authorized status. When creating an object to inject, we then create it as the correct sub-class and call its `Authorize` function, which is free to then operate as it needs to. You can even reuse this structure for authorization of other things elsewhere too:

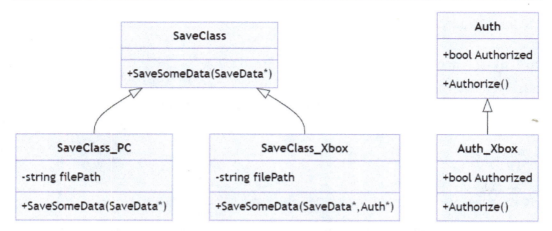

Figure 2.6 – UML diagram showing SaveClass alongside Auth hierarchy

So, if we apply dependency inversion to our code, it would look like:

Excerpt of a program that implements Figure 2.6

```
Auth* _Authorizer = new Auth_Xbox();
SaveClass* _XboxSaver = new SaveClass_Xbox("filePath");
SaveData* _DataToSave;

//Some code to prepare save data

_Authorizer->Authorize();
_XboxSaver->SaveSomeData(_DataToSave, _Authorizer);
```

That's a lot of theory to have gone through, but with it, we can make tangible improvements to dire systems. These improvements should shore up the foundations of our code, setting it up to receive the more structure-heavy patterns we will be using later in this book.

The next section will guide you through an average beginner developer's process of creating a project and getting something working quickly so that you can then apply what you have learned to see a tangible improvement.

Exploring solutions to common problems

We are going to explore a series of examples of Blueprint scripts where the intended result has been achieved but causes issues simply due to the approach taken. We will then offer an improved approach, which makes the Blueprint more performant, tidier, more reusable, or easier to expand at a later point in development. The purpose of this section is to help you begin to recognize potential areas for improvement within your own code, regardless of whether you are working in Blueprint or C++.

> **Important note**
>
> Everything we are about to look at in this section is bad code to prove a point. Even the fixes that follow are not perfect, but it has been written for the purpose of education and so is simplified somewhat.

For this section, you will need an empty project that you may have already created at the beginning of the chapter and the `content` folder from the `chapter2` branch of the GitHub link provided at the start of the chapter.

If you've not used GitHub before, click the **Code** button on the branch view and then click **Download Zip**. This will download a `.zip` folder that you can then extract into your new project folder.

Place the `HelloPatterns` folder from `.zip` file directly into your content directory; this will provide you with some examples of bad practises that we will first review before explaining how to fix them.

To fix the problems, first, duplicate each of the Blueprint assets and change the suffix from `_Bad` to `_Better`. This will ensure you can look back and see the differences between where we start and where we finish when reworking these Blueprints.

To start fixing these Blueprints, you will need to open them by double-clicking on them in the **Content Browser**. Once open, navigate to the **Event Graph** tab where you will find the examples we've covered. Follow the steps under each of the **Solution** headings to improve the Blueprints.

The moving box problem

The first example we're going to look at is a simple moving box (a static mesh component) that slides from a start location (relative `0,0,50`) to an end location (relative `200,0,50`). Both locations have been stored as vector variables within the Blueprint and their defaults set, as mentioned.

> **Note**
>
> Here, we are using relative location, a vector variable that defines the location of an entity based on the location of its parent. A relative location of (0,0,50) on a component whose parent actor is positioned at (0,0,0) in the world would also be (0,0,50) in the world. However, if the parent actor were to move to (10,10,0), the component's world location would be (10,10,50) as the world and relative locations are combined to provide a world location for the component.

The Blueprint code to move the box has two logic chains selected via a branch node (Blueprint's version of an `if` statement) using **Event Tick** (Unreal's form of the `Update` loop). One logic chain is used to move the box forward, from start to end, and the other is used to move it back. The decision of which to use is controlled by the **Forward** Boolean variable:

Figure 2.7 – The Event Tick Branch for selecting which direction to move

The *move forward* logic chain adds 1 cm to the *x* coordinate of the box's relative location and updates the position of the box, relative to the actor using a **Set Relative Location** node.

A check then takes place, comparing the current relative location to the end point with a 0.0 error tolerance. If the locations are equal, then **Forward** is set to **False**, which means on the next tick, the branch will select the reverse logic change:

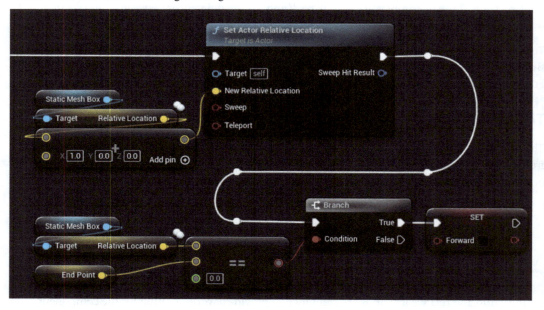

Figure 2.8 – The move forward logic chain for the moving box

The *move backward* logic chain subtracts 1 cm from the *x* coordinate of the box's relative location and updates the position in the same way. The difference here is that the check compares the relative location to the end point and, when they are equal, sets **Forward** to **True**, flicking the branch to the *move forward* logic on the next tick. This repeats indefinitely or until the actor is destroyed:

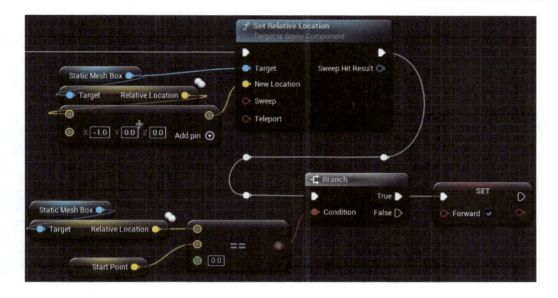

Figure 2.9 – The move backward logic chain for moving the box

The issue we have here is we are checking on every tick if the box has arrived at its destination. Doing the comparison on one box may not cause many issues, but if you are moving a lot of items in a scene, this sort of innocent calculation is a waste of resources, something we are keen to avoid.

Solution to the moving box problem

Let's take a look at building a better approach using a timeline:

1. Begin by deleting all of the nodes in the **Event Graph**, except for the **Set Relative Location** and **Static Mesh Box** nodes.

2. Start by adding a custom event node by right-clicking on the graph, typing custom event, and pressing *Enter*. You can select the **Add Custom Event…** option at any point; typing more of the name will simply help reduce the options available, hopefully speeding things up:

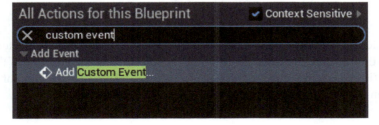

Figure 2.10 – Node creation in the Event Graph with custom event filtered

3. Call the custom event `PingPongMovement`—this is intended at this point to be quite literal as that's what we are coding:

Figure 2.11 – A custom event node for PingPongMovement

4. Drag out of the **exec pin** (white arrow) and type `add timeline`, select **Add Timeline...** from the popup, and call it `T_MoveBox` when prompted. This will give you a timeline node, a special type of node that contains one or more graphs that can be used to provide values for other nodes that are called during the timeline node's **Update** chain:

Figure 2.12 – A timeline node, named T_MoveBox

5. Double-click the **T_MoveBox** node; this will open a new tab to allow us to create and edit the graphs in this timeline. Note that the length of the timeline, shown at the top of the tab, is 5.0 seconds. This means the timeline will execute its update chain for a duration of 5.0 seconds from when it is first called:

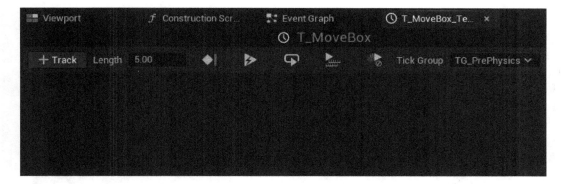

Figure 2.13 – An empty timeline tab with default settings

6. Click the +**Track** button and choose **Add Float Track**.

7. Name the track `MovementAlpha`. This will add a float track graph and float value output to the timeline node:

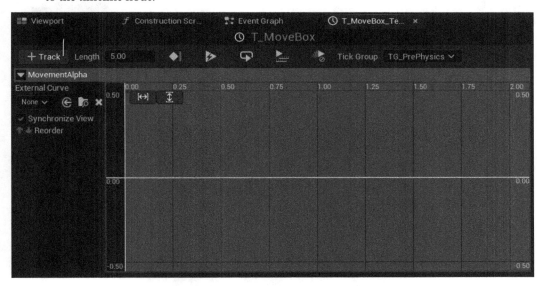

Figure 2.14 – A float track added to the timeline

8. Right-click on the float graph and select **Add Key to CurveFloat_0**; repeat this to create a second key.

9. Select the first key and, using the **Time** and **Value** variable boxes at the top of the graph, enter `0.0` into both variables.

10. Select the second key and set **Time** to 5 . 0 and **Value** to 1 . 0:

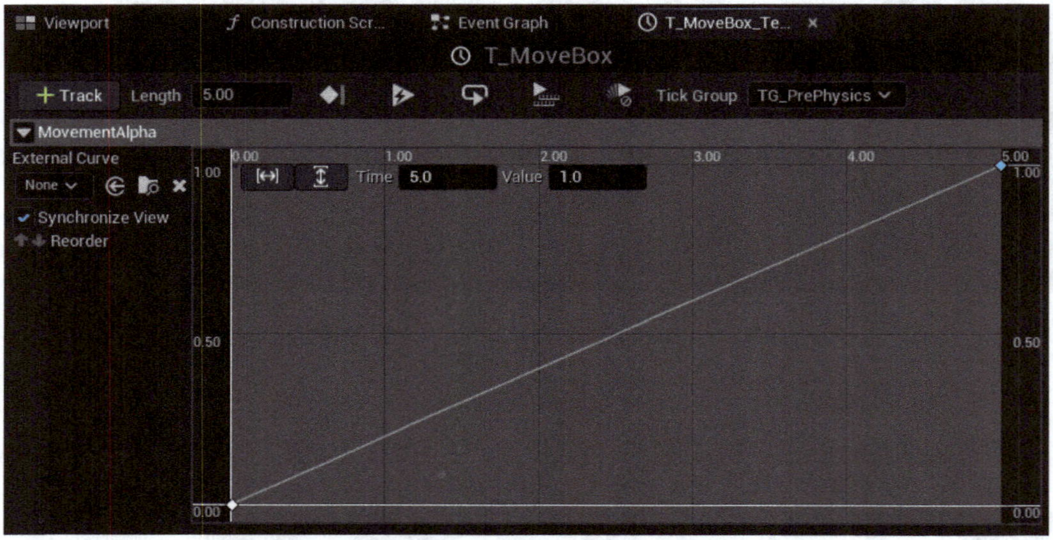

Figure 2.15 – The MovementAlpha track with linear graph

11. Return to the **Event Graph**; you should now see the timeline node has a **Movement Alpha** float pin on the right-hand side of the node; accessing this pin will provide you with the current value of the MovementAlpha curve based on the current time of the timeline:

Figure 2.16 – The timeline node, now with the Movement Alpha float pin

12. Position and connect the **Set Relative Location** and **Static Mesh Box** nodes from the old solution to the **Update** output pin from the timeline.

13. Drag from the **New Location** pin of the **Set Relative Location** node and type `lerp` to create a **Lerp (Vector)** node. This setup allows us to interpolate between two vectors, to control where our box moves, relative to the object's location.

14. Drag and drop the **Start Point** variable from the left side of the **Blueprint Editor** window onto the **A** pin of the **Lerp (Vector)** node. The **A** pin is the start value for the lerp.

15. Drag and drop the **End Point** variable onto the **B** pin of the **Lerp (Vector)** node. The **B** pin is the end value for the lerp.

16. Connect the **Alpha** pin of the **Lerp (Vector)** node to the **Movement Alpha** pin of the timeline node:

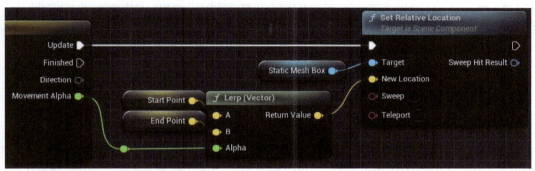

Figure 2.17 – The update logic of the timeline completed

Now the update logic is complete, we need to set our Blueprint up to start the timeline. To do this, proceed as follows:

17. Right-click on the **Event Graph** and create an `Event BeginPlay` node. This is a built-in event that is called when an actor is created (either at the beginning of the game or when spawned by another function).

18. Drag out of the **Event BeginPlay** node and type `Ping Pong Movement`; you should now see a **Ping Pong Movement** option under the **Call Function** rollout—select it. This will create a call to our custom event:

Figure 2.18 – Ping Pong Movement being called from the Event BeginPlay event

This will now move the box from the start location to the end location in relative space; give it a test by placing the BP_MovingBox_Better Blueprint from the **Content Browser** into the world and clicking the *Play* button above the viewport:

Figure 2.19 – Play in Editor controls from the main Unreal Engine Editor user interface

Next, we want to set the Blueprint up so that the box moves back to the start location once it has reached the end location and loops indefinitely. To do this, proceed as follows:

1. Detach the **Ping Pong Movement** custom event node from the timeline node by holding *Alt* on the keyboard and clicking either end of the connection.

2. Drag from the **Ping Pong Movement** custom event node and create a Flip Flop node. This node swaps between exec pins **A** and **B** each time it is entered. The node always starts with **A** for its first run. The **bool** pin can be used to inform other logic, but we won't need that for this solution.

3. Connect the **A** pin of the **Flip Flop** node to the **Play from Start** pin of the **T_MoveBox** timeline node.

4. Connect the **B** pin of the **Flip Flop** node to the **Reverse from End** pin of the **T_MoveBox** timeline node:

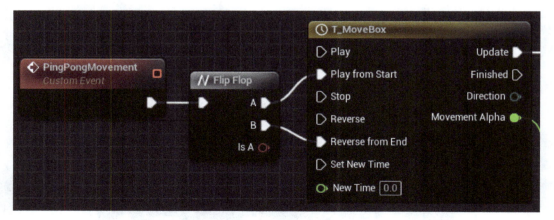

Figure 2.20 – Flip Flop node and timeline

5. Lastly, connect the **Finished** pin of the timeline node to a **Ping Pong Movement** function call node (either the existing one or a new one), which will restart the process every time it finishes:

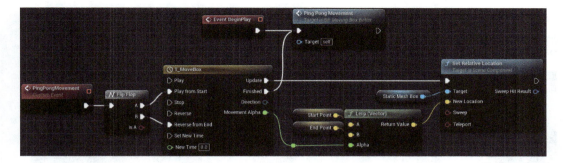

Figure 2.21 – The improved Blueprint for the moving box, laid out for readability in the editor

Now, try pressing the *Play* button in the editor (the *Play* button above the viewport) again. You should now see that the box moves in one direction and then the other, over and over.

This approach offers a more performant solution due to not requiring the comparison. The update logic is still constantly running, in the same way as the tick approach was for the **Set Relative Location** node, but because the timeline controls when the direction change occurs, we no longer need to do any comparisons of vector locations after every move.

The timeline can also afford us more control over situations when we want to move things. The current setup features linear movement controlled by the linear curve. Changing the key types to automatic (right-click the key) or adding additional keys can provide more interesting movements and allow you to incorporate considerations such as the principles of animation into your moving objects, something that is very difficult to do without using a curve.

The rotating box problem

The second example we're going to look at is a simple rotating box. The Blueprint actor contains the same static mesh component, but this time it's set up to spin in place 2 degrees of rotation on each tick:

Figure 2.22 – The rotating box Blueprint code

The Blueprint code also does a check for when the rotation of the box gets to 360 degrees or above and minuses 360 from the rotation value, to ensure that we keep the rotation value from spiraling out of control as the game plays:

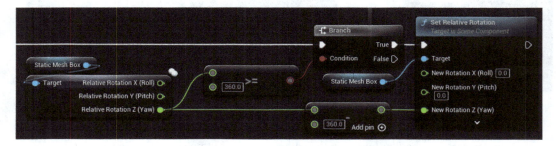

Figure 2.23 – The check to maintain rotations between 0 and 360 degrees

The check that's been put in place is relatively sensible, and if it were part of a single rotation or working with a character rotation within an animation Blueprint, then it would be ideal.

The problem with this implementation is that we are once again working on tick and doing checks that will become more cumbersome with every additional rotating box.

Solution to the rotating box problem

For the rotating box, the solution is to use a pre-existing solution. Instead of coding differently, we can use a component that is provided as part of UE5 to achieve the desired result. Follow the next steps:

1. Firstly, delete all of the logic from the **Event Graph**, it is not required.

2. In the **Components** tab, click the **Add** button and search for a **Rotating Movement** component:

Figure 2.24 – Rotating Movement component selected in the Add components list

3. Create a new variable by clicking the + in the **Variables** rollout of the **My Blueprint** tab.

4. Name the variable Rotation Rate and set the type to Rotator by clicking on the current type (typically **Boolean** if you are working with a new Blueprint) and selecting **Rotator** from the list.

5. Make the variable **Instance Editable** by clicking the *eye* icon to the right of the variable type—this will allow the user to set the rotation rate of each box in the scene.

6. Compile the Blueprint and set the default value of the **Rotation Rate** variable to 0,0,180:

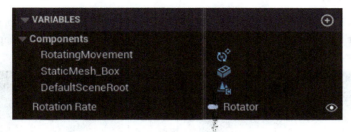

Figure 2.25 – Resulting variables list in the My Blueprint tab

7. In the **Construction Script**, drag the **Rotating Movement** component into the graph from the **Components** list. From the resulting **Rotating Movement** node, drag out and create a Set Rotation Rate node. Connect the **Rotation Rate** variable you just created to the **Rotation Rate** pin:

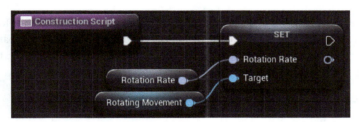

Figure 2.26 – Construction script logic, setting the Rotating Movement
component's Rotation Rate using the created variable

Now, when placing BP_RotatingBox_Better into the world, you will be able to set the **Rotation Rate** value in the **Defaults** section of the **Detail** panel. Try placing multiple boxes and setting different values. When you press the *Play* button in **Editor**, you should see the boxes rotating as desired.

This solution offers a more performant approach again due to its lack of reliance on **Event Tick** as well as the removal of the comparison checking when the rotation exceeds 360 degrees in order to maintain values inside the 0-360 range.

Another benefit of this approach is that we now have an **Instance Editable** variable for rotation rate, which offers much more control and customization to the user, allowing each box to have its own rotation rate. The variable also allows us to spin the box on all three axes, whereas the original approach only rotated the box around the *z* (yaw) axis.

The cascading cast chain problem

This example is something we see quite regularly with new developers when communicating between different Blueprint classes.

The implementation here is for a game where a player character could be carrying one of three weapons: a pistol, shotgun, or rifle. Casts have been used to identify the class of the carried weapon, and when the cast returns true, each weapon's fire event is called:

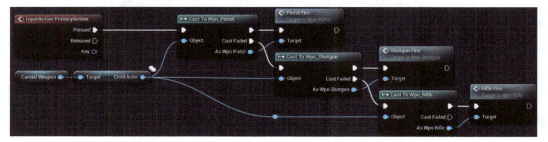

Figure 2.27 – A cascading cast chain in the Blueprint in the CH_Example_Bad asset

There are a few issues here. Firstly, it is the nature of the cast node in Unreal. While fine for prototyping, casts carry a resource impact where each actor we attempt to cast to gets loaded as part of the actor. So, in this case, all three weapons are included with the character in order to be able to check if the **Carried Weapon** child actor matches their classes.

You can see the effect of casts on memory by checking the size map of any actor.

To do this, proceed as follows:

1. Select the **Actor** Blueprint in the **Content Browser**.
2. Right-click and select **Size Map…**.

This will open a popup that will show the size (in memory) of the chosen actor:

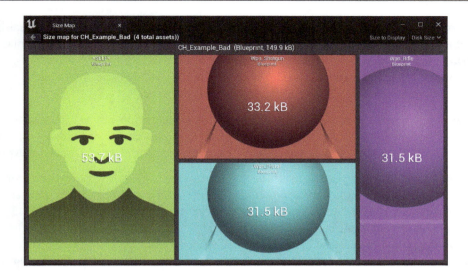

Figure 2.28 – Size map of the character Blueprint class with a cascading cast chain

The resulting Blueprint (which only has the basic character class elements and the cast chain) is 149.9 kB, compared to the 56.1 kB of a standard character with just the child actor set as a pistol. This could be made smaller, further optimizing the Blueprint's memory impact, by setting the default class of the child actor to be actor, but this isn't necessary to prove the point we are trying to make; casting causes unnecessary memory issues:

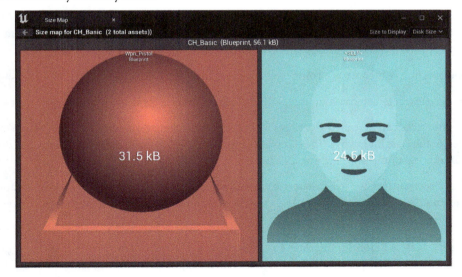

Figure 2.29 – Size map of a character Blueprint class with just a child actor component

In both examples shown previously, the `Wpn_Pistol` class is just an empty actor with a custom event that would, in an eventual game, fire the weapon; however the class currently only contains a **Print String** node with the name of the weapon followed by the word `Fire`. If you consider adding meshes, textures, particle systems, and audio components, the potential size of a character with a cascading cast chain becomes exponentially larger than it needs to be.

The second issue is the processing time of casting and waiting for failure before casting to the next weapon class and potentially waiting for that to fail. While we are talking milliseconds, it all adds up, especially if there are multiple characters and/or your non-player characters use the same character Blueprint.

The final issue with the cascading cast chain is the inflexibility (or extra work) that this approach provides. With the current setup, only the three weapons included in the chain can ever be used by the character. To expand the number of available weapons (as the project progresses or as part of a post-release piece of downloadable content), more casts will need to be added to the chain, multiplying the effects of the two aforementioned issues as well as being generally time-consuming to add them, particularly if there is any other logic required as part of the chain such as ammo management and cooldowns.

If you want to see the current setup's output, then open the `TestLevel` map from the `TestFiles` directory, click *Play*, and then click the left mouse button; you will see a print in the top-left corner of the screen related to each weapon. Pressing the *1*, *2*, and *3* keys on your keyboard will swap between different weapon classes.

Let's take a look at the solution to this problem.

Solution to the cascading chain problem

There are two potential solutions to the problem; the first is to use a parent and child class approach, casting to the parent class, which will allow **Event Fire** to be called on all of the children. As this still uses a cast, the parent class will still be included in the character, inflating the memory usage.

The second solution, which we are going to implement, is using an interface.

Interfaces allow two actors to communicate, without the need to identify the class type of the target actor.

Blueprint interfaces can be called from any Blueprint graph using an actor reference (which is the highest step of the class hierarchy). The receiving actor is required to implement the interface to define how it will respond to the interface event call.

In building this solution, we will create a simple Blueprint interface, add it to all of the weapons, and call it from the character, simply referencing the weapon as an actor class, negating the need for casting. Proceed as follows:

1. Firstly, create a Blueprint interface by right-clicking in the **Content Browser** and navigating to **Blueprints | Blueprint Interface**. Name the resulting asset `BI_Weapon`. We use the prefix `BI` to label this as a Blueprint interface.

2. Open the **BI_Weapon** asset by double-clicking it. You should see that in the top-right **My Blueprint** panel, a function has been created for you called **NewFunction_0**, and its name is currently set up to be edited. Rename this `Fire`.

3. Open **WPN_Pistol**, click **Class Settings** in the toolbar, and then click the **Add** dropdown and search for `BI_Weapon`. Clicking it will add the **BI Weapon** interface to the **Implemented Interfaces** list:

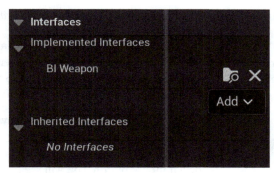

Figure 2.30 – BI Weapon in the Implemented Interfaces list

4. This will now have added a **Fire** function under the **Interfaces** rollout in the **My Blueprint** tab. Double-click it; this will create an **Event Fire** interface event:

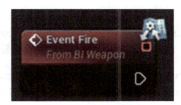

Figure 2.31 – The Event Fire interface event, denoted by the interface icon over the top-right corner

5. Drag out from the **Event Fire** node and add a `Print String` node, then replace **Hello** with `Pistol - Interface Fire`.

6. Repeat *steps 3 to 5* on both the **Wpn_Rifle** and **Wpn_Shotgun** assets, changing the **Print String** node to include the name of each weapon.

7. Duplicate the **CH_Example_Bad** asset and rename it **CH_Example_Better**.

8. Open **CH_Example_Better** and delete all of the nodes from the **InputAction PrimaryAction** logic flow, except for the **Carried Weapon** and **Target Child Actor** nodes.

9. Drag from the **Child Actor** node and search for `Fire (Message)`; this will create a **Fire** interface event call, which will be called on whichever class is currently set as the **Carried Weapon** class:

Figure 2.32 – The resulting InputAction PrimaryAction Blueprint logic with the interface call

In order to test this approach, we need to swap the **Default Pawn Class** in our game mode (the class that governs default game elements such as pawn and player controller, as well as handling game-level variables and events) for the TestLevel map to the better example we've just created.

10. Open GM_Test from the TestFiles directory in the **Content Browser**.

11. Change the **Default Pawn Class** dropdown to **CH_Example_Better**.

Now, when you play the level and click the left mouse button, you will see the new interface prints.

The outcome we have achieved is essentially the same as before; however, the code used to achieve it is tidier, faster, uses less memory, and is infinitely expandable by avoiding casts and using the interface.

The trade-off

With any implementation of a pattern or fix, there is a trade-off between what you gain and what it costs, and it's important to consider the long-term effects when implementing patterns.

In many of the cases we are exploring, the gain is simplicity, readability, or a reduction in memory footprint, which should help your game to run smoothly. These are all key elements worth maintaining a good code base for. The cost is often time. Some patterns may take longer to implement, and, in some cases, having to refactor code to work in a specific way can use up valuable time.

The long-term effects, however, outweigh the initial time cost as the time saved later in development to build on top of or into existing systems will prove beneficial later in your game's development cycle.

Summary

In this chapter, we discussed the S.O.L.I.D. principles that underpin good code, exploring specific examples and how, as they expand in complexity, we need to consider these principles to keep the code functioning. These principles are widely recognized across the game development industry, and so understanding them will not only help improve code efficiency and readability but also allow smooth communication with co-developers on larger projects.

We also explored a series of common problems that new Unreal Engine users encounter and the solutions to them, looking specifically at a series of common tasks such as moving items and managing a selection of weapons on a character. The solutions, while specific, offer insights into how easily a developer can find themselves with large, lumbering Blueprints that can be solved by utilizing built-in systems and tools.

In the next chapter we will be taking these principles of clean code and seeing how Epic Games have applied them through the engine when we look at some of the patterns they have built for us. The utility of the engine as it were. These patterns will include double buffer, flyweight and spatial partitioning.

UE5 Patterns in Action – Double Buffer, Flyweight, and Spatial Partitioning

Hopefully, you will have realized by now that Unreal Engine 5 is a really big engine. Behind the scenes, it already employs a lot of the patterns that we will cover in later chapters. This chapter will break down the **double buffer**, **flyweight**, and **spatial partitioning** patterns. You don't need to build these three patterns yourself as Unreal already has good implementations, but knowledge of their existence and how they have been created will help you build on top of them. This chapter will look into how Unreal implements each pattern into a system and what problems they are solving for you in the process. This should give you a roadmap to not only discover more about the engine but also some examples of good practice to reference moving forward.

In this chapter, we're going to cover the following main topics:

- Double buffer
- Flyweight
- Spatial partitioning

Technical requirements

In this chapter, we are going to delve into three patterns that are important to how any commercial engine performs. There will be some use of **Big O notation**, which is simply a low-resolution way of measuring the time efficiency of an algorithm. The lower the resulting number when replacing the n with a large number, such as 1,000, the better the time efficiency. For example, an algorithm that compares each element of an array with every other element of the same array could be described as $O(n^2)$. This comes from the idea that the algorithm is a couple of nested for loops that run for the length of the input data. Maybe then we improve efficiency, meaning we don't need to recheck elements as we go through making the seconds for the loop shorter with each iteration. This would result in O(n log2n). Looking at these values, you can tell that for large numbers, O(n^2) is far worse, giving an estimated cost of 1,000,000 executions for an array of size 1,000, whereas the same array put into O(n log2n) costs only 9,965 executions.

Later in the chapter, we will once again make use of some example files. You can grab these from the *Chapter 3* branch on GitHub at `https://github.com/PacktPublishing/Game-Development-Patterns-with-Unreal-Engine-5/tree/main/Chapter03`

Double buffer

For this pattern, we need to imagine a photocopier being operated by an artist. The artist has been commissioned to deliver at least two copies of every picture they draw, so they think smart and use a photocopier to copy their work. To save time moving artwork from the easel to the copier, they simply put their canvas straight onto the scanner. They then paint as fast as they can and hit *Copy* at the same time. What follows is a race where the artist needs to paint each line fast enough to stay ahead of the scanning head. If they are successful, the artwork and the copy will look the same with no extra time taken. More than likely, the scanning head will get in front of the artist, which will result in a picture up until the point where the artist fell behind and a blank copy after that point. This is known as **frame tearing**, the issue we are trying to solve. Frame tearing occurs when the frame buffer, our example artist's canvas, has not been fully updated by the time the read pointer passes over it to draw to the display. This was a big problem back in the early days of graphics, where you only have 76 clock cycles to load the next row of pixels into the frame buffer before it was dispatched to the display. This perpetual race with the GPU, which immediately restarts once it finishes, creates a problem for graphics programmers. *Figure 3.1* shows how the artifacts created due to frame tearing can cause issues for the person viewing the screen:

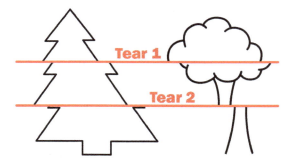

Figure 3.1 – Image showing visual artifacts created by screen tearing in two places

If this issue occurs frequently in a game being played at a standard 60 frames per second, it can cause everything from dissatisfaction with the game to nausea.

Figure 3.2 shows what visually happens when the read pointer (red) starts behind the write pointer (green): in the middle column, the write pointer is overtaken by the read pointer, meaning from this point on, none of the work done by the write pointer will make it to the screen as the read pointer will have already read that pixel data. This will result in the screen looking like the right column, even though all the data was rendered to the buffer:

Screen

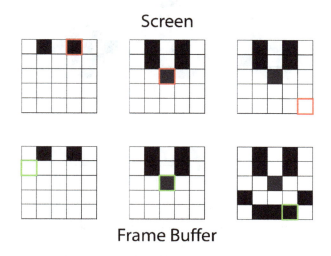

Frame Buffer

Figure 3.2 – Diagram showing how race conditions between the read
(red) and write (green) pointers cause frame tearing

This raises the question: how can we calculate all the shaders required to draw every visible object in this strict time limit?

The answer is to not try to. Instead of optimizing to beat the current situation, we can change the situation. Currently, the display is catching the write pointer mid-calculation; instead, we could draw our picture to a different buffer, a back buffer. This allows the GPU to calculate in peace with no fear that the half-finished image will be shoved in the face of an unsuspecting viewer. Once the new frame is calculated into the back buffer, a pointer changes to make this new picture the new active frame buffer. The next time the GPU needs to send data to the screen, we know the image is finished. In *Figure 3.3*, we can see that although the read pointer overtakes the write pointer in the middle column, because it is pulling from a buffer that is complete, no tear is rendered to the screen:

Screen

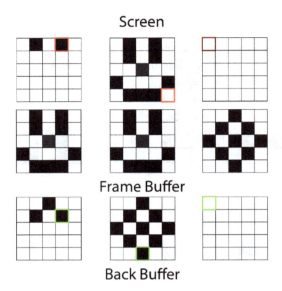

Frame Buffer

Back Buffer

Figure 3.3 – Diagram showing the screen (top) rendering from the frame buffer
(middle) while the write pointer works on the back buffer (bottom)

In slow calculations, this means that an old image will be displayed on the screen for maybe multiple calls, as the new image can't be completed in good time, but it is better to have this stutter than a tear across the middle of the screen.

If the opposite was true and we well outpaced the display frequency of the attached monitor, then we have a lot of wasted time where the GPU simply waits. The backup artwork is ready to ship and the main one has not been taken yet. Most gamers know this as a setting called **vertical synchronization** (**VSync**). With VSync off, as soon as the back buffer is full, the back and frame buffers are swapped, which can lead to situations where one frame has a sudden drop in performance and there is no back buffer sitting in reserve to fill the visual gap. This does lead to more responsiveness as the image being shown is technically more recent, but the difference in latency between having the setting on and off is negligible in comparison to the visual issue of a screen tear.

Something we can do to increase our productivity during large VSync wait times is to use more buffers. This idea of a double buffer extends to infinity. Triple buffers would be efficient to implement if there is a chance that low refresh rate monitors will be used with high-end graphics cards. This way, once the first back buffer is full of data, the GPU can turn to another back buffer so as not to remain idle. When the display call comes through, the newest finished buffer becomes the frame buffer and all others, bar the one that is currently being filled, are wiped and sent to the back of the queue. This strategy is largely unnecessary with modern hardware, though, which is why Unreal only uses a double-buffered system.

It is not so easy to find the exact place where this pattern is used within the Unreal Engine source code, as it has a different implementation for each render pipeline. By default, Unreal launches with **DirectX**. DirectX is a graphics pipeline designed for use on the Windows platform by Microsoft and has built-in methods for creating and managing frame buffers. DirectX calls the double buffer process the **swap chain**. This particular implementation can be found by navigating to your Unreal Engine 5 install folder and heading to the following path:

```
<UE_5.0 folder>\Engine\Source\Runtime\Windows\D3D11RHI\
```

This folder contains all the DirectX-specific files in the public and private folders. This includes files such as D3D11Viewport.cpp, which contains the editor implementation of a swap chain of buffers. These files are very dense DirectX code, and so won't be covered in great detail here, but this is a good starting point to explore how Unreal deals with its graphics pipeline. Keep in mind that Unreal Engine also supports **Vulkan**, and this has a different implementation.

At this point, you may be considering double buffers as a low-level tool. Primarily, they are, but you can use them for other high-level gameplay situations. Any time that data gathering and processing has been parallelized to run on multiple threads, a double buffer could be used to make sure processing is being done on cohesive data. This could take the form of heatmap analytics, keeping track of where players die on a level, or AI gathering data about its surroundings to make decisions.

Let's move from a system running under your code all the time to an automatic process within the engine that silently boosts your game efficiency: the flyweight pattern.

Flyweight

The flyweight pattern is focused on reducing memory usage for large collections of objects and reducing the time spent loading them in. For the flyweight pattern, we first need to consider three things:

- **Intrinsic data** – The data values that are immutable and considered to always be true on an object
- **Extrinsic data** – The data values that are mutable and considered to be changeable per instance
- **Memory costs** – All data must be stored somewhere and for loaded objects, that means on our RAM

With that in mind, let's look at trees in the forest. Taking a simple approach, we could load in and store the model, texture, and transform data once for each tree. This would make our data storage look like *Figure 3.4*, with each tree connected to its own plot of memory, holding its copy of the data needed for rendering:

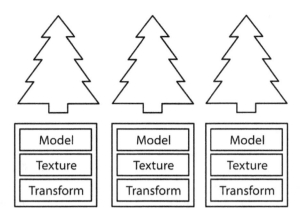

Figure 3.4 – Diagram showing the data associated with each tree

If we consider the memory cost of a **model** and **texture** combined to be about 5 MB and the transform component to be 12 bytes of data per tree, then our memory footprint for three trees is a little over 15 MB. This may not seem like much, but it scales badly. Considering a forest of 100 trees, we would need a little over 500 MB of RAM to hold what is essentially set dressing. Something must be done, so let's consider what data in the concept of a tree can be considered intrinsic, and what needs to remain extrinsic to each tree. As all our trees are identical to look at, all the data for visuals (model and texture data) can probably be safely made intrinsic. Great, we have a definite set of data that is each type; so what? Well, the flyweight pattern places all data we consider to be intrinsic in shared memory. In practice, this looks like *Figure 3.5*, where all intrinsic visual data that will be shared between all trees has been collected into a shared block and you can spatially visualize the difference in memory cost. Things like the objects transform still remain extrinsic, as each tree will at least have a different position vector:

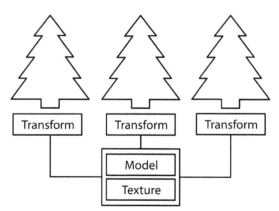

Figure 3.5 – Diagram showing how the separation of intrinsic data
saves memory with repeated objects spatially

In terms of memory allocation, the implemented flyweight pattern has a fixed cost of 5 MB for the shared model and texture data and a variable cost of 12 bytes per tree. This means our forest of 100 trees now only takes up 5.1 MB, which is a far cry from the 500 MB we needed previously.

With the basics down, let's investigate how our system might break. Well, we obviously get a benefit from sharing data, but what if we wanted two types of trees in our forest with different model and texture data? You'd think, based on the previous explanation, that this would make that data extrinsic and therefore not eligible for the flyweight pattern. However, you can invoke the type object pattern, which we will discuss in greater detail in a later chapter. The type object pattern collects intrinsic data about a *type* and applies the same idea. This calculation for memory cost would then look like the amount of types you have multiplied by 5 MB, plus 12 bytes per object in the world. The type object pattern, of course, has some other benefits that we will look at in *Chapter 9*, but this is how it pairs with the flyweight pattern to give an expandable, efficient system.

Unreal Engine applies the flyweight pattern by default to every asset loaded into a project. Assets are then referenced from a single place when needed and their data is shared across all references. If you need to make a change to a specific loaded version of an asset, you need to actively break the flyweight pattern to do so. An example of where this may be needed is in dynamic material instances. Some value may be needed to alter the appearance of only one object in the scene, so a copy of the intrinsic data for the material must be made.

The last pattern for this chapter is less automatic and silent than the previous two. It is a subsystem designed around making vast worlds possible and performant to play in. In Unreal Engine, we call it **World Partition** (based on the toolset available in the editor to manage how a large-scale world is split up), but more generally it is known as the spatial partitioning pattern.

Spatial partitioning

Imagine the Matrix is a real concept, and we are living in a giant physics simulation. Anybody who has dabbled in physics simulations knows that you don't need many interacting objects to make the simulation chug. The naïve solution is to check every object against every other object leading to an $O(n^2-n)$ solution, where you can see in *Figure 3.6* that 4 objects have 12 collision checks:

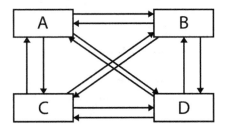

Figure 3.6 – Diagram showing each collision check performed in an inefficient collision detection solution

Improvements can obviously be made to not repeat calculations that have already been made, bringing us down to $O(nLog2(n))$ where those same four points use only six collision checks. This is better, but there is only so far we can go by checking every object against every other object. We could just not calculate some of them, but that would be like giving some people in our Matrix simulation wall hacks, allowing them to possibly phase through walls, or worse, each other.

We could check the distance between objects before we check their collisions. This would mean objects far apart would not even try to collide. This may work for more complex collision geometries but realistically, if we are smart developers using only primitives for collision, we would just be adding to all the calculations. This would in the worst case lower the efficiency to $O(nLog2(n) + n)$. The theory has merit, though, so instead of checking every object we should instead group them based on geography. The world needs to be divided into cells; we'll call them chunks. Each cell knows the objects it fully encompasses and its position in the grid of cells. We need only then to check collisions between objects in neighboring cells. This is the basis of spatial partitioning.

What if an object is bigger than the cell it is centered on? We can group cells together into a bigger chunk and include any objects that cross the cell boundaries that the new cell group fully encompasses. This process is then repeated until the cell contains the whole world. This creates a tree structure of cells. To then check an object's collision, only objects in the tree directly connected above and below need to be checked. This has the potential to cut out most checks we would have otherwise done.

One issue is how we establish where each object is in the tree. This does have some overhead cost, meaning for small simulations, it is wild overkill. That said, if you are going to use this across your entire world and only update the tree positions of moving objects, you end up with massive efficiency gains. 2D implementations might follow a structure called a **quadtree** (pictured in *Figure 3.7*):

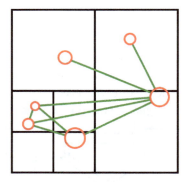

Figure 3.7 – Diagram showing how a quadtree can reduce the number of needed collision checks (green) to eight for these six objects (red) when they are in these positions

Here, you can clearly see how objects in different cells (or quads) do not check collision with each other. As an object crosses a boundary, it starts existing in the cell a layer above and must check collision with all objects in child cells. Generally, this is not the case, and so with a large number of objects, there will always be a performance improvement.

Unreal Engine 5 uses the idea of a quadtree in three dimensions called an **octree**. It is the same principle of dividing space in half along each axis, but because we are dealing with three axes, each cubic area divides into eight cells, hence the name octree. The octree in Unreal resides in the `Fscene` class and is used by the render thread to determine quickly whether an object exists in the light's area of effect. This way, the number of lighting calculations is drastically reduced:

```
ScenePrivate.h ×

3026        /** An octree containing the shadow-casting local lights in the scene. */
3027        FSceneLightOctree LocalShadowCastingLightOctree;
```

Figure 3.8 – Screenshot of the octree location in the ScenePrivate.h file

> **Extra note**
>
> If you drill down through the struct and class definitions from the line pictured in *Figure 3.8*, you will eventually find a `Toctree2` class name (the first one has been deprecated). In this class, you can explore Epic's implementation of the octree.

That is not all. With Unreal Engine 5 comes the introduction of a system called World Partition. This is another implementation of a spatial partition, this time in two dimensions, that is used for a different purpose. As a developer, you must define what size cell Unreal should use, and unlike the tree variants, this is limited to a single layer of depth. That being said, it does provide a fast way of hooking up large worlds with **level of detail** (**LOD**) systems to affect everything from visuals to collision meshes. LOD systems are used to reduce how much detail is being rendered (in either a 3D mesh or texture) when the object being rendered appears smaller on the screen. So, an object further away from the screen will feature an LOD mesh that has a much lower triangle count than if the same object was drawn directly in front of the player.

Let's take a look at exploring World Partition by looking at an example map we've provided that uses an Open World level. We will begin by looking at how the example works, then change some variables to get different results to better understand how the system works. The example levels used for this section are available in the `PatternsInAction` project on GitHub that is linked at the start of the chapter. Unlike in *Chapter 2*, this time we have provided a full Unreal Project, and this time, you can extract the `.zip` contents and double-click on the `PatternsInAction.uproject` file.

Introducing World Partition

Firstly, open the `WorldPartitionExample` map file from the `Content\Maps` folder. This map should open by default when you launch the project but in case it doesn't, navigate to **File | Open Level** and select it from the directories. You should then see a world with a series of blueprint actors, all of which are displayed as primitive shapes:

Figure 3.9 – World Partition example map from the PatternsinAction project

This level already has World Partition enabled and set up for the terrain. This is standard for the Open World level template.

Let's start by opening the World Partition setting tool so we can see how it works. If it hasn't opened by default, go to **Window** | **World Partition** | **World Partition Editor**:

Figure 3.10 – Accessing the World Partition Editor from the menus

This should open **World Partition Editor** in the same section of the user interface as the **Details** tab; you can zoom in using your scroll wheel, after which you will eventually be able to see the grid reference system for the map (provided you have turned on **Show Cell Coords** in the top menu of the **World Partition** editor). It's worth noting that this grid does not directly relate to the cell grid produced by the World Partition system. Instead, it offers a grid based on an editor cell size which, by default, is set to 12,800 (128 square meters), half the size of the default World Partition cell, which is 25,600 (256 square meters):

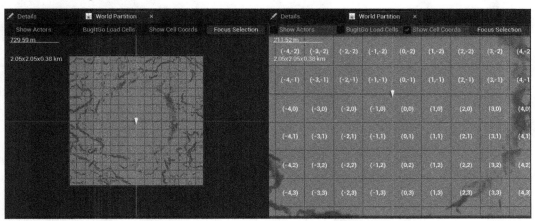

Figure 3.11 – World Partition editor tab, resized horizontally so the text on the left is visible (left) and then zoomed in with Show Grid Coords ON to show the grid reference system (right)

World Partition splits (partitions) the world based on a cell grid, with each cell corresponding to a physical section of the level. If an actor's bounding box is in that cell, it will be drawn (or hidden) depending on whether the cell is within a set distance of a **streaming source** (an entity in the world that controls when World Partition cells are loaded), such as the player character or an actor with a World Partition streaming source component. This allows Unreal to divide the map and only draw the required cells, making it possible for developers to build incredibly large maps.

To determine the distance from a streaming source that a cell must be within to be drawn, the engine uses the **Loading Range** variable, which can be found in **World Settings | World Partition | Runtime Settings | Grids | Index [0]**. The value of **Loading Range** sets the radius from the character that a cell needs to be within for it to be loaded and rendered. The default value for **Loading Range** is 76,800, which is shown in **Unreal Units** (uu). 1 uu = 1 cm, so the default radius is the in-game equivalent of 768 m:

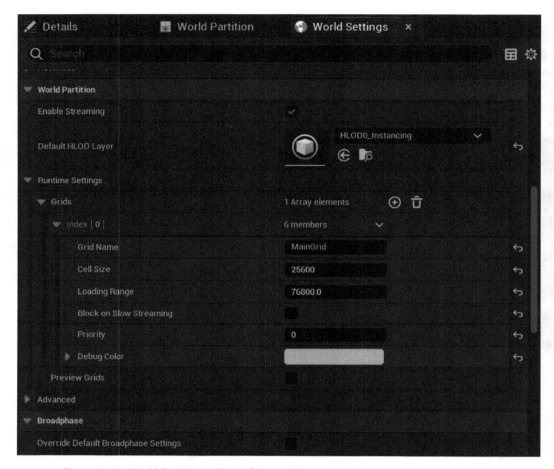

Figure 3.12 – World Partition rollout of the World Settings tab showing the grid settings

The **Index [0]** section of **World Settings** also contains variables to determine the size of each cell on the grid. In a built-up city environment, where the visible distance is shortened due to the height of buildings, you could set a smaller **Cell Size** value and **Loading Range** value compared to an open landscape, such as a desert.

While it may appear logical to significantly reduce the cell size and loading range, setting these too low will result in frame rate issues owing to a bottleneck caused by trying to load (and unload) too many cells at one time.

Now that we've had a look at how the cell grid is made up, we can start to look at how that cell grid controls when objects load and unload based on the position of streaming sources.

Understanding when objects are loaded

To get a better understanding of how World Partition works, we can change some of the **Runtime Settings** values in **Index [0]**. We will make changes to **Loading Range** and **Cell Size** to control the way World Partition works within our example map. The changes we make will adjust when the blueprints in the map are loaded and unloaded. Observing the differences will help you to experience World Partition (and spatial partitioning) in action.

Before we make any changes, press the *Play* button to play the level in the editor viewport. You should notice a stream of print strings on the left side of the screen. This is because each of the blueprint actors have been set up to print a short message and their display name.

Move around the map using typical first-person *W*, *A*, *S*, and *D* keys to control the character. You should notice that, if you move all the way to one side of the map, prints change as fewer objects are printing their messages.

Once you've seen this effect, try changing **Loading Range** from 76800 to 768, making the radius that objects need to be within to load 100 times smaller than the default settings. This will cause a significantly different result. Now, try testing the level once again. You should notice that only two objects are visible: a tall rectangular box and a cone, both of which are in front of the character from the starting position. These are visible because a small section of each mesh is inside a partition cell that is within a 7.68 m radius of the player character.

Walking toward either object will eventually cause other objects to suddenly appear (you will need to get very close) as you get within range of the neighboring partition cells.

To further understand why we get this behavior, let's change the **World Partition Editor Cell Size** variable:

1. Below the **Runtime Settings** section we've used so far, expand the **Advanced** section by clicking the title or the arrow to the side.

2. Set the **World Partition Editor Cell Size** to 25600 to match the **Cell Size** value from the **Index [0]** settings, ensuring our editor display matches the behavior we are experiencing when we play the level.

3. Once you have made the change, rebuild the Minimap by navigating to **Build | Build Minimap** from the top menu. You will be asked to save the level and any modified assets, and then you will be shown a series of progress bars while the build takes place.

4. Switch back to the **World Partition** editor tab; you should now see a lot fewer cells in the grid, as shown in *Figure 3.13*:

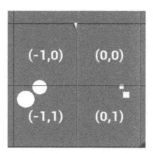

Figure 3.13 – Updated editor cell display showing how the visible objects
are partially inside the initially loaded cells (-1,0) and (0,0)

The grid should now show where each of the actors is positioned, with at least one of each of the object types (cone and cube) in a grid cell that the character starts within (or very close to). The second object of each pair sits in the neighboring grid cell, so will only appear once the player character is within 7.68 m of the cell, which is a very short distance when you consider each cell is 256 m x 256 m and the default loading range that we changed was originally a radius of 768 m. Based on the default values, when the character stands at the starting location at the origin of the grid, 36 of the cells are being loaded as opposed to the 4 that are loaded with the shortened range of 7.68 m, which would be inappropriate for use in a game but works as an extreme example of how the system works:

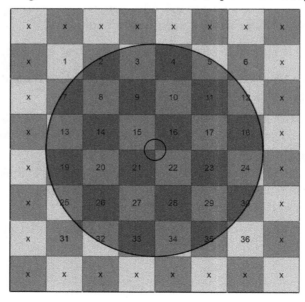

Figure 3.14 – Visual representation of the overlapping area of a 768 m radius circle over 256 m x 256 m grid cells, resulting in 36 cells being loaded compared to a radius of 7.68 m (inner circle) only overlapping 4

Now that we've set an extreme value, try some different values, such as setting **Loading Range** to 38400, which will result in more objects being loaded at any one point. Standing at the center of the map will load a similar number of visible objects to the default setting, but you will see the objects disappear much sooner as you walk away from them. Remember, you can always return to the default value of 76800.

This shows how, with different loading ranges, we can control when parts of our 3D world appear. Getting the correct values for your game might require some testing and, in some cases, the inclusion of scalability settings to ensure your game runs on different hardware. For now, the key thing is understanding how to control when things load and unload based on streaming sources so that once you have a 3D world to control, you can make the most of World Partition.

Ensuring actors are loaded when required

There are two ways to ensure that actors are loaded and ready for the player. One option is to force an actor to always be loaded by preventing it from being loaded and unloaded by World Partition. This could be useful if you have managers in the world who are required to control aspects across the entire level, or if you have a monument that needs to be always visible. To do this, do the following:

1. Select the chosen actor in the level.

2. In the **Details** panel, scroll down to the **World Partition** section:

Figure 3.15 – World Partition settings in the Details panel with an actor selected

3. Set **Is Spatially Loaded** to OFF.

The second approach is to include actors in the world that have a World Partition streaming source component.

We've included a BP_StreamingSourceActor blueprint, which has a World Partition streaming source component set up with the target grid already set up. To see it in action, do the following:

1. Drag a BP_StreamingSourceActor blueprint from the Content Browser/ Blueprints folder into the world between the two cone actors.

2. Set **Loading Range** in **World Settings** back to a small value such as 768 for testing purposes.

When testing the level, you should now see that, unlike when we first set the loading range to 768, both cone objects are visible.

You will, however, notice that after five seconds of gameplay, one of the cones (the furthest one from the player) will disappear. This is because the BP_StreamingSourceActor blueprint includes some crude code to toggle the streaming source on and off. This provides a nice insight into how a streaming source actor can be used to control loading cells ready for the player, in cases such as where the character will be teleported to a new location that needs to be fully loaded before the player is relocated.

In *Figure 3.16* you can see the logic within the BP_StreamingSourceActor's event graph.

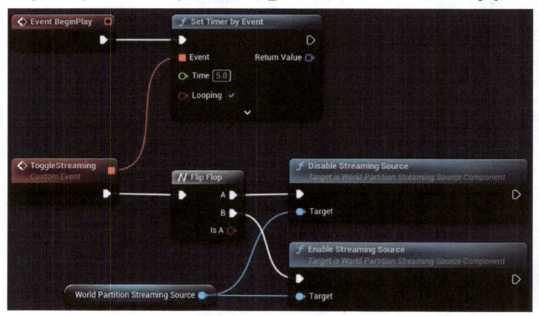

Figure 3.16 – Crude blueprint setup that disables and enables the
streaming source every five seconds using a timer

Now that we've explored World Partition, we will look at using it with levels that have been created without it.

Enabling World Partition on an existing level

To convert an existing level (that wasn't created using the Open World starter map) to make use of World Partition, all you need to do is use the built-in tool:

1. Ensure the level you want to convert to a World Partition-enabled level is saved.
2. Navigate to **Tools | Convert Level** from the top menu.
3. From the popup, select the map you would like to convert.
4. Click **OK**.

The tool will create a new version of the map with the _WP suffix in the name unless you select **In Place** from the final settings dialog, which will simply overwrite your existing level.

Summary

In this chapter, we have discussed a number of existing patterns that are present in Unreal Engine, focusing on double buffer, flyweight, and spatial partitioning.

We have taken a look at some of these systems, including exploring editor examples of spatial partitioning using the World Partition system, experimenting with different variables to control when objects are loaded and unloaded, and testing an actor with a World Partition streaming source component.

In the next chapter, we are going to explore some pre-made patterns in Unreal. We will be looking at creating components using the update method, and creating a simple AI-controlled **Real-Time Strategy** (**RTS**) game unit using Unreal's AI **Behavior Trees**.

4

Premade Patterns in UE5 – Component, Update Method, and Behavior Tree

This chapter will focus on the three main patterns widely used in game development that Unreal Engine 5 offers robust support and tools for, including pre-built implementations and editors – namely the **component**, **update method**, and **behavior tree** patterns. We will discuss the theory of why they exist and explore how you can implement them with guided exercises in our custom framework.

Understanding the tools at your disposal will improve your development speed, saving you from reinventing the proverbial wheel. Even if these are tools you are familiar with in Blueprint, some insight into the C++ workings will hopefully improve your effectiveness wherever you use them.

In this chapter, we'll be covering the following topics:

- Understanding and creating components
- Applying the update method for prototyping gameplay
- Working with behavior trees

Technical requirements

For this chapter, you will need to download the starter RTS framework from the *Chapter 4* branch of the GitHub repository, which can be downloaded from `https://github.com/PacktPublishing/Game-Development-Patterns-with-Unreal-Engine-5/tree/main/Chapter04`

We will be building elements for this framework in the following chapters, building up a series of gameplay features using patterns as we explore them.

Understanding and creating components

One of the first things you learn when programming is to try to never repeat yourself. In fact, every technique you learn, from loops to functions to class encapsulation, is focused on reusing code with less typing. Building up the analogy, a loop reuses lines of code in one area so that you don't need to repeat them next to each other. A function reuses blocks of code so that you don't need to repeat them across your class. Then a class lets you reuse sets of functions and data in instances, so you don't need to repeat logic across your program.

How does this help? Well, in games, any object that can be seen probably has some rendering element allowing it to be drawn to the screen. The code for rendering your object in your chosen graphics pipeline follows a standardized structure and is likely to be the same across every object that needs to be rendered. Even thinking about this possible repetition should be ringing alarm bells. Initially, it is a waste of your time to write out the same lines multiple times in different places. Then, if you need to make a structural change later down the line, you suddenly have a scavenger hunt trying to find all the places it needs to be changed.

There is a better way.

Enter, the component pattern. Essentially, we utilize the reusable aspect of classes as described previously to encapsulate all logic around a repeatable behavior. This forms a sort of building block template that we can instance and reference in many places. When the time comes to update the logic, it exists in only one place and will affect all areas that reference it.

Practically, this also makes our code around this functionality cleaner. Components are generally designed to be completely self-contained, which means that our collision code will deal with context for us and the rendering component will deal with linking model and texture assets together with materials. This leads to fewer checks needed before calling functions on these components and reduces unintended behavior.

For instance, when dealing with collision checks between two objects, you don't need to find out what shape collider objects A and B have before checking if they can collide; the code for deciphering that is contained in the collider component itself. You also only get collision signals from the collider component; it doesn't make your model flash when it collides with something. Using components effectively can lead to a modular code architecture that will be key to getting the most out of a large engine such as Unreal with a team of developers.

> **Important note**
> Notice how this links back to the single responsibility principle from our SOLID principles of code back in *Chapter 2*. Good patterns almost always reflect these principles. In this case, our components do one thing each that makes our code modular, and therefore, more useful.

Unreal Engine 5 is based on what is called an **entity component system** (**ECS**). This essentially means that all objects in the game world are considered entities. Each entity comprises components that, as we have established, encapsulate some behavior. When you open an actor's **Blueprint Editor** and look at its hierarchy, you are looking at the components that make up that entity (or in this case, actor). Take, for example, the default character in the framework adjacent to this book. We have left the hierarchy, which is all we will be focusing on for now, as mostly the same as the example character that Epic provides in their sample projects. BP_TopDownCharacter3 has the following components:

- **Capsule** – set as the root; it describes the shape of the object for the physics engine
- **Mesh** – holds all the data for how to render the object to the screen
- **Spring arm** – dynamic component for holding objects at a distance
- **Camera** – attached to the spring arm, this describes the position and settings of the camera to render the screen

Components vary in importance and size. The preceding list shows how a spring arm component, which is relatively small and unimportant, can be used with a larger, more important component such as the camera within the same actor. Size doesn't dictate importance though, so let's have a look at the first component you may create to make your development a touch easier – a health component. Our aim is to make a simple-to-use tool for designers to drop onto any actor, giving them the ability to track a health value, which is attached to the built-in damage system. The component should also provide hooks for notifying when damage has been taken and when the health value hits zero.

With the example framework open, click on **Tools | New C++ Class…**.

You will be presented with the following window, showing all the parent classes you can extend:

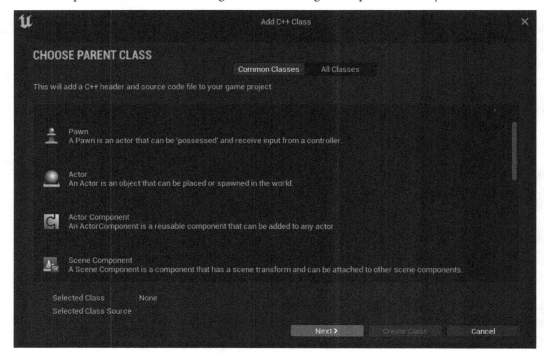

Figure 4.1 – Window for choosing a class to inherit from

We are interested in **Actor, Component** so select that and click on **Next**.

In the following window, we will name the component something sensible such as `HealthComponent`:

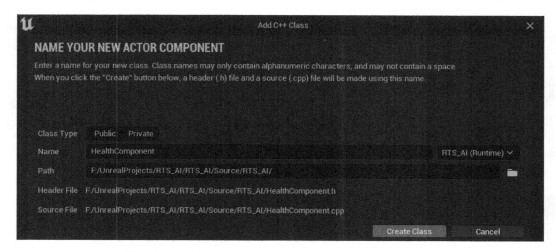

Figure 4.2 – Window for naming and scoping a new C++ class

The rest of the settings don't matter for this example, but they do impact the scope of our component. This is only something you need to be concerned with if you are building large projects with multiple sub-systems that may need to cross-reference classes. For now, let the settings be as is and click on **Create Class**. This will launch the IDE Unreal is linked to, which by default is Microsoft Visual Studio.

> **Important note**
>
> While tools such as Visual Studio are great IDEs, at the time of writing, there is none better than JetBrains Rider for Unreal development. As of the 2022 version, Unreal-specific structures are built into their code auto-completer and error checker. However, this software is not free, so if you are using Visual Studio, then note that some of your code may display errors, but it will build and run fine.

With the class open, you should have a header (.h) and a body (.cpp) file. We will start by discussing requirements drawn from our task aim and building our header. Our component needs to do the following:

- Contain a value for health
- Be able to receive a damage signal
- Notify when taking damage and when dead

Tackling our value for health is easy: we only need to define a float variable to track the current value and another to allow setting of the maximum value from the editor. With that in mind, setting the property specifiers in the UPROPERTY block is the most important part. Taking damage is also quite simple as Unreal Engine 5 has a damage interface system that proliferates every AActor. To hook into this, all you need is a function that shares a signature with the OnTakeAnyDamage event, as shown in the following HealthComponent.h code snippet. Lastly, to notify the owning AActor that the component has taken damage or is out of health, we will use some dynamic multicast delegates. These will be covered more in *Chapter 6* when we look into clean communication, but for now, think of them as smoke signals coming from the component that the owning AActor can watch for.

So, let's write the code in the header file:

HealthComponent.h

```
#pragma once

#include "CoreMinimal.h"
#include "Components/ActorComponent.h"
#include "HealthComponent.generated.h"

DECLARE_DYNAMIC_MULTICAST_DELEGATE_OneParam(
    FComponentDeadSignature, AController*, causer);
DECLARE_DYNAMIC_MULTICAST_DELEGATE_ThreeParams(
```

```cpp
    FComponentDamagedSignature, AController*, causer,
    float, damage, float, newHealth);

UCLASS( ClassGroup=(Custom), meta=(BlueprintSpawnableComponent) )
class RTS_AI_API UHealthComponent : public UActorComponent
{
    GENERATED_BODY()

public:
    UHealthComponent();

    UPROPERTY(BlueprintAssignable)
    FComponentDeadSignature onComponentDead;
    UPROPERTY(BlueprintAssignable)
    FComponentDamagedSignature onComponentDamaged;

protected:
    UPROPERTY(BlueprintReadOnly, VisibleAnywhere)
    float _currentHealth;
    UPROPERTY(BlueprintReadWrite, EditAnywhere)
    float _maxHealth;

    virtual void BeginPlay() override;

    UFUNCTION()
    void DamageTaken(AActor* damagedActor, float damage,
        const UDamageType* damageType, AController*
        instigator, AActor* causer);
};
```

Next, we need to fill out the function definitions for the constructor, `BeginPlay` and `DamageTaken`. The constructor is simple enough as it just sets advised values for designers. Here, we only set the max health. `BeginPlay` is a little more interesting. Shown in the following code block, `BeginPlay` is where we link our damage function to the event in the owning `AActor`. `DamageTaken` does all the work in this component, but it is equally simple, with two lines being taken for "Broadcast"-ing the delegates.

Important note

We will be looking in more depth at what events do and what some of the terminology means later, in *Chapter 6*. For now, the important aspect is that this component does function the way we need it to.

Let's add those to the body file:

HealthComponent.cpp

```
UHealthComponent::UHealthComponent()
{
    _maxHealth = 100.f;
}

void UHealthComponent::BeginPlay()
{
Super::BeginPlay();
    GetOwner()->OnTakeAnyDamage.AddDynamic(this,
        &UHealthComponent::DamageTaken);
    _currentHealth = _maxHealth;
}

void UHealthComponent::DamageTaken(AActor* damagedActor,
    float damage, const UDamageType* damageType,
    AController* instigator, AActor* causer)
{
    _currentHealth=FMath::Max(_currentHealth-damage, 0.f);
    onComponentDamaged.Broadcast(instigator,damage, _currentHealth);
    if(_currentHealth <= 0.f)
    {
        onComponentDead.Broadcast(instigator);
    }
}
```

To use this component, head into a Blueprint class and click the **Add Component** button in the Blueprint Editor window under **Components**. Searching here for the name of your component without the word `Component` should find it. Then, after adding it, you can link logic into it by clicking the green plus buttons next to the events we made in C++:

Figure 4.3 – Delegates created in C++ showing in the Blueprint editor window

As a final note on this pattern, you may also have heard the term ECS bandied around with games such as *Doom Eternal* and *Overwatch*, but this is slightly different from Unreal's implementation. The ECS being spoken about there is what we call *data-oriented* ECS. This means that components store only data. The systems are the functions and are stored separately. Then the systems run on archetypes of components we call entities. Entities are stored contiguously in memory, which makes processing systems across large arrays of entities much faster and even possible to easily multi-thread.

The trade-off with this is that, while in principle it sounds simple, in practice it is a paradigm shift in how to approach programming as large as the jump from the linear *hello world* style programs with everything in one file to object-oriented projects with many classes and complex internal structures. This, combined with the infancy of the approach, means it is not practical to develop a large game and expect the entire programming team to hit the ground running with this system. If you would like to learn more about full data-oriented ECS, then there are plenty of resources on the topic. The keywords for proper research into the topic would be data-oriented, archetypes, and ECS, in some order, but due to the low adoption rate of the technique, resources are generally either terse or surface-level. The concept is best learned practically by pulling the ENTT C++ library apart to see how it implements the pattern. This is best for practical learners as it is a raw implementation with no other distractions.

Components are everywhere in modern game design, and now you should have some understanding of how to structure your code with them. Another ubiquitous tool in your developer belt is the update method.

Applying the update method for prototyping gameplay

The update method is all about abstraction. Let's learn through an example. Imagine building a *Pong* clone with only rudimentary knowledge of C++. It might look something like this:

Naïve Pong code

```
Entity aiPaddle;
Entity ball;
Vector2D direction;

// Main game loop
while (looping)
{
    Vector2D ballPos = ball.GetPosition();
    ballPos += direction;
    ball.SetPosition(ballPos);

    if (ballPos.y > aiPaddle.GetPosition().y)
    {
        aiPaddle.MoveUp();
```

```
    }
    else
    {
        aiPaddle.MoveDown();
    }

    // Input, collisions, and rendering...
}
```

This code may work for *Pong* but the problem is pretty clear. As you add more and more types of objects to your game, the ways you have to deal with every frame balloons your code. This is not sustainable.

The solution can be found in the minds of every person with an overbearing micro-managing boss: "*Just let me do my job.*" In this situation, the main loop is the manager, and the entities are sitting on the sidelines. They exist but are largely having logic run at them rather than on them. Why not trust these experts and make each entity responsible for the logic under its namesake? We keep a list of references in the main loop, but we forget who each of them are and resort to referencing them as numbers, or indexes in an array.

This new impersonal approach frees up the main loop to go through all entities and call a shared `Update()` method on them all. Each entity can now deal with its own world within its own override of the `Update` function. The following code snippet is an example of how you might manage this new approach:

Better Pong code

```
Entity entities[];

//Main game loop
while(looping)
{
    for(Entity e : entities)
    {
        e.Update();
    }
//Collision and rendering...
}
```

Unreal Engine's implementation of `Update` is called "`Tick`." It fires once every rendered frame and allows developers to provide each actor they make with a different behavior. There is, however, a big issue with the update method pattern, which is why we will be exploring it in more depth in *Chapter 5*.

Now that we've covered the update method pattern, we are going to move on to the implementation of the behavior tree pattern in Unreal Engine 5 by exploring the behavior tree system inside the Unreal Engine editor.

Working with behavior trees

The last pattern for this chapter is a little specific but no less useful than the others. Behavior trees define decision trees for AI brains to run. These trees define how information stored about the situation the AI thinks it is in will affect the actions it takes. Instead of compartmentalizing code into discrete blocks like other AI brain solutions, the behavior tree favors fragmentation with the aim of reuse. This flows back to the idea from the beginning of the chapter about the reuse of code for efficiency. A tree may look complex with many branches and leaves (yes, that is what we call the composites and tasks that make up a behavior tree), but on closer inspection, you will see that there are only a few building blocks and their arrangement is what creates the illusion of complex decision making.

The flow through a simple tree starts at the root and evaluates each node in the layer below in ascending order from 0 (order matters here) until it gets a success response. At this point, the tree knows it has found the correct task for its situation and can reset. As mentioned before, there are two types of simple nodes: composite and task. Composite nodes direct the flow of logic with their own sets of rules, whilst tasks are the actions and can succeed or fail depending on the logic contained within them (either in Blueprint or C++ code), and can never have a layer below them. The behavior tree executes each node in order until they reach the end of the list, or a node fails. They succeed when all their children succeed.

We will be using selector and sequence nodes, both of which are types of composite nodes. A selector node executes all of its children from left to right; it stops executing when one of its children succeeds. A sequence node also executes all of its children from left to right; however, unlike the selector node, it stops executing when one of its children fails. There are other more specialized nodes, but these two composite nodes are the foundations for any system you could wish to design.

The other composite node type, which we won't be using, is the **simple parallel** node type. The simple parallel node allows you to run two child nodes at the same time: one must be a task, and the other can contain a complete sub-tree or branch of the behavior tree.

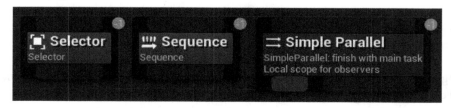

Figure 4.4 – The three types of composite nodes: Selector, Sequence, and Simple Parallel

A notable use of behavior trees in games is *Halo 2*. As mentioned above, the order of the nodes in the tree matters and can radically change how AI behaves as it will tend to favor branches higher in the order. To differentiate the varied enemy types, Microsoft designed different behavior trees for each one. *Grunts* have a higher-order branch for fleeing so they tend to seem more scared, whereas *Elites* can re-order their own trees based on what the player is doing to make them seem adaptable. Notably, if you get in a vehicle enemy, *Elites* will raise the priority of them getting in a vehicle as vehicle-to-vehicle combat is more fun than the alternative and vice versa. There are many more tricks going on in that game, but this is not the place to continue, so back to UE5.

Let's look at building a behavior tree for a simple **real-time strategy** (**RTS**) AI in the RTS framework. We will introduce the various building blocks of behavior trees, including selectors, sequences, decorators, services, and tasks.

For this implementation, we are going to use Blueprint to create the service (a collection of functions called by a behavior tree node) and tasks as opposed to using C++, as the main focus is on the implementation of the behavior tree pattern in Unreal Engine 5.

The framework contains a series of assets that have already been set up for you. Any Blueprint logic that we've created has been commented, if you would like to explore them further. Inside the `Content/RTS/Blueprints` folder, you will find a series of unreal assets used to create the basic elements of our RTS example game:

- **GM_RTS**: This is the game mode. It contains references to the various classes for the framework to work.

- **PC_RTS**: This is the player controller. It contains the functionality to left-click selectable units and right-click on the world in order to tell the unit where to go.

- **BP_CameraPawn**: This is our pawn. It is a simple pawn blueprint with a camera that is set up to be placed directly in the center of the map.

- **BPI_Units**: This is our Blueprint interface. It contains three functions: **SetMoveLocation**, **Stop**, and **AttackTarget**:

  - **SetMoveLocation** has a vector input, **Target Location**

  - **AttackTarget** has an object input, **TargetToAttack**

  - **Stop** is just a function name and does not have any inputs or outputs

- **BP_EliteUnit**: This is our character blueprint. It is the AI character that we are going to be building upon.

  This Blueprint extends from the Unreal base character class and implements the `BPI_Units` Blueprint interface.

The skeletal mesh component has been set up to use a green-tinted material instance placed on the default mesh, which is inherited from the base character class. We are making use of the standard mannequin character mesh here.

This Blueprint currently contains no functionality.

- **BP_EnemyUnit**: This is a second character blueprint. This is a simple character blueprint with no functionality.

  The skeletal mesh component has been set up with a red-tinted material instance so that it is clearly identifiable.

So, we are going to build an AI unit that the player will be able to control. The player will only be able to tell the unit where to move. They will do this by first selecting the unit using a left-click and then selecting a desired location for the unit to move to by right-clicking on the floor.

The unit will then move to the location. When not moving, the unit will search to see if any enemy units are within a radius of its current location. If it finds an enemy, the unit will then turn to face the enemy and shoot at it (using a simple line trace for now).

To achieve this, we need to create a series of assets. We are going to create an AI controller asset, for which we will create a Blackboard asset that contains the variables used inside the behavior tree.

Creating the AI controller

The first thing we need to create is an AI controller Blueprint asset. This asset allows us to replace the human controller for a pawn with an AI solution using a behavior tree. For this, we need to do the following:

1. Inside the Unreal Engine Editor, open the **Content Browser** and navigate to the Content/ RTS/Blueprints folder.
2. Right-click within the folder in the **Content Browser** and create a new **Blueprint Class**.
3. Expand the **ALL CLASSES** rollout and search for AIController.
4. Select the **AIController** entry from the list and click **Select**.

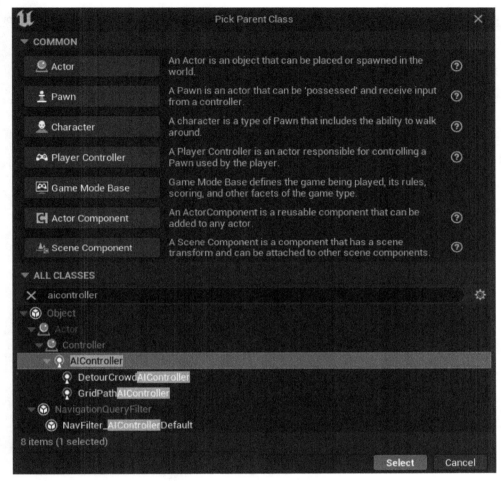

Figure 4.5 – The Pick Parent Class dialog with the AIController parent class
selected as the chosen parent class for the new Blueprint

5.   Name the new Blueprint AIC_EliteUnit.

Now we have our new AI controller, we now need to set it as the default AI controller for the *Elite*
unit. This will ensure that whenever an Elite unit is spawned or placed within the game, it will have
a controller. To assign the default AI controller, we need to do the following:

1.   From the Blueprints folder, open the **BP_EliteUnit** Blueprint asset.

2.   Click the **Class Defaults** button in the top menu.

3.   In the **Details** panel, set the **AI Controller Class** to AIC_EliteUnit.

Now we have an AI controller, we can start building our AI system. Before we can create the behavior tree, we first need to set up a **Blackboard** asset.

Creating the Blackboard asset

Blackboards are used to define variables (known as keys) and store their values, which will allow our behavior tree to make decisions.

To create the Blackboard asset, we need to do the following:

1. In the **Content Browser** window, right-click and select **Artificial Intelligence | Blackboard**.

Figure 4.6 – The Artificial Intelligence options when creating a new asset

2. Name the new **Blackboard Asset** BB_EliteUnit.

3. Open the BB_EliteUnit Blackboard asset.

4. Add two new keys and set the **Key Type** as shown:

    *   MoveToLocation: Vector

    *   TargetEntity: Object

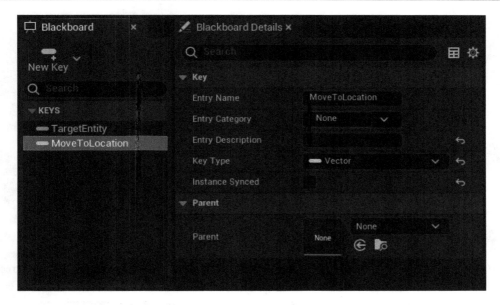

Figure 4.7 – The Blackboard asset with two keys

Now that we have our Blackboard setup with our variables (keys), we can go ahead and create the behavior tree asset and get building our AI system.

Building the behavior tree

Now we have our Blackboard, we can create our behavior tree asset and link the main assets together:

1. In the **Content Browser** window, right-click and select **Artificial Intelligence | Behavior Tree**.

2. Name the new **Behavior Tree** asset BT_EliteUnit.

3. Open the BT_EliteUnit asset.

4. Set the **Blackboard Asset** setting in the **Details** panel to **BB_EliteUnit**.

Figure 4.8 – The Blackboard Asset setting in the Details panel of a Behavior Tree

5. Open the `AIC_EliteUnit` Blueprint asset (the AI controller) and activate the behavior tree when the unit is possessed using **Event On Possess**:

Figure 4.9 – Blueprint nodes to run the Behavior Tree when the unit is possessed

With that done, we now have created the behavior tree asset; we can start building the AI system.

Basic movement branch

To start building the brains of our system, we are going to set up the following:

- A selector node, which is used to branch out into other sections of the tree
- A sequence node, which allows us to move through multiple tasks, in order, from left to right
- Our first task, which will use the existing **Move To** task

This will enable our character to be moved around the world. This part of the system requires no additional logic, just setting up within the behavior tree itself.

So, let's start building:

1. Open the BT_EliteUnit behavior tree.

2. Drag from the dark gray section at the bottom of the root node and create a **Selector** node by selecting it from the **Composites** section of the popup; name this RootSelector.

Figure 4.10 – Options available when dragging from the root node in the behavior tree

3. From the bottom of the new **Selector** node, drag out again and add a **Sequence** node. Name this node MoveToTargetLocation.

4. For one final time, drag from the bottom of the new **Sequence** node and add **Move To** from the **Tasks** section of the selection popup:

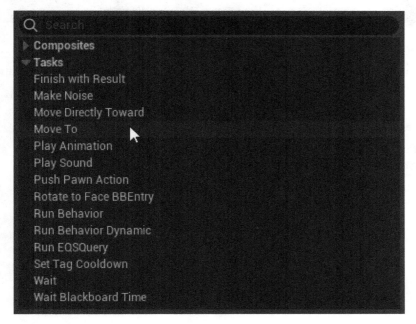

Figure 4.11 – Behavior tree creation popup with Move To highlighted

5. Set the Blackboard key to MoveToLocation – this is the vector key in the Blackboard asset.

At the moment, this behavior tree will run; however, the task (*Move To*) will try to move the character to 0,0,0 because we have yet to set a value for the **MoveToLocation** key on the Blackboard.

The Blackboard keys can be set from multiple places if you can provide a reference to the owner of the Blackboard.

We've already provided the connection between the player controller Blueprint and the BP_EliteUnit character. Now we can set up the character to do something with those values:

1. Open the BP_EliteUnit character Blueprint.

2. Create a new Event Possessed logic chain that casts **New Controller** to AIC_EliteUnit and then stores the reference as a variable called AIController. Do this by creating the Blueprint nodes as shown (to create the **Set AIController** node, right-click on the **As AIC Elite Unit** pin and select **Promote to Variable.** Call the variable AIController):

Figure 4.12 – Event Possessed logic on the BP_EliteUnit character blueprint

3. From the **Interfaces** section of the **My Blueprint** tab, double-click the **Set Move Location** label. This will create an **Event Set Move Location** node.

4. From the **Event Set Move Location** node, promote the **TargetLocation** vector to a variable. Call the variable MoveToLocation.

5. Drag the **AIController** variable into the graph, add a Get Blackboard node, and from that, add a Set Value as Vector node.

6. From the **Key Name** pin, drag out and add a Make Literal Name node. Set **Value** to MoveToLocation – this is the key name on the Blackboard.

7. Use the **MoveToLocation** vector variable as the **Vector Value** input.

Figure 4.13 – Event Set Move Location logic on the BP_EliteUnit character Blueprint

This logic will, when the event is called, pass **Target Location** through to the behavior tree by setting the MoveToLocation key value.

8. Repeat this process for the **Stop** interface function from the **Interfaces** section of the **My Blueprint** tab, except instead of using a pin on the event to set the **MoveToLocation** vector, use a **Get Actor Location** node.

Figure 4.14 – Event Stop logic on the BP_EliteUnit character Blueprint

The last thing left to do is tell the **Move To** task on the behavior tree to observe changes to the MoveToLocation key on the Blackboard:

1. Select the **Move To** node on the behavior tree.

2. In the **Details** panel, check the **Observe Blackboard Value** box.

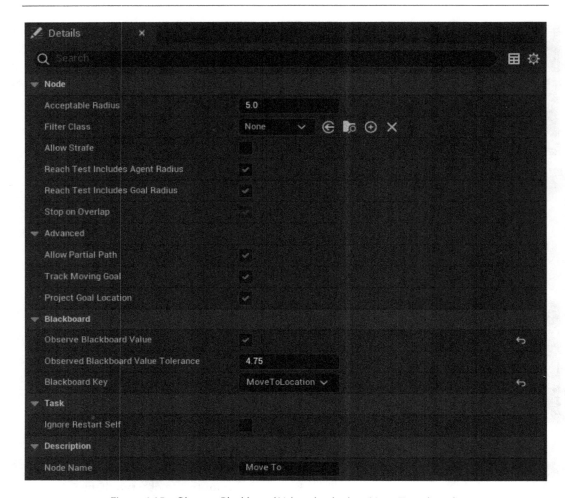

Figure 4.15 – Observe Blackboard Value checked on Move To task node

Now that the basic movement is set up, the character can be instructed to move around the world. Next, we are going to add a branch to our tree that will focus on the shooting part of our system.

Shooting branch and task

The next element of the system we are going to build is a means to enable the character to shoot enemies. For this, we are going to need to create a new task to build into the tree:

1. To start with, click the **New Task** button at the top of the **Behavior Tree** editor and select **BTTask_BlueprintBase** from the list.

Figure 4.16 – New Task dropdown with BTTask_BlueprintBase selected

2. From the **Save Asset As** pop-up window, ensure the Blueprints folder is selected and set the name as BTT_ShootTarget.

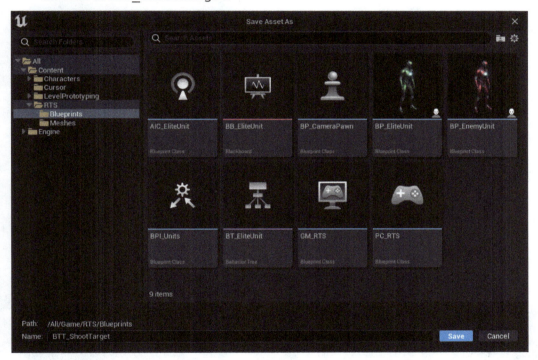

Figure 4.17 – Save Asset As window for behavior tree tasks

This has now created a behavior tree task asset, which should open automatically. If it doesn't, you will find it as a Blueprint asset in the **Content Browser**.

With the task created, we now need to add some logic for what the task will do.

A task works by utilizing the **Event Receive Execute AI** node, which provides references to the owning controller and the controlled pawn. This allows us to communicate easily with the pawn. In this case, we are going to call the interface event `AttackTarget`, which we will set up to just print a `Shoot` message for now.

Behavior tree tasks need any logic chains to end with a **Finish Execute** node. The node has one input pin, a Boolean for `Success`. If the logic has completed the task, then this should be set to **True**, if it hasn't, it should be set to **False**. For this example, we will only have a **True** result as all we are doing is calling an event on the character Blueprint.

3.   In the `BTT_ShootTarget` Blueprint, add an **Event Receive Execute AI** node either by right-clicking in the viewport and searching for it or by clicking the **Override** dropdown that appears on the **Functions** rollout title when you hover over it.

4.   Drag from the **Controlled Pawn** pin and call the `AttackTarget` interface event.

5.   From the **Target to Attack** input pin, drag out and add a `Get Blackboard Value as Object` node. This will allow us to get the value of an object key.

6.   Right-click the **Key** input pin, select **Promote to Variable**, and name the new variable `TargetEntity` – it is vital that this matches the name of the key on the Blackboard.

7.   Set the `TargetEntity` variable to be **Instance Editable** by clicking the eye to the right of the variable type in the **Variables** list.

Figure 4.18 – Instance Editable set to True on the TargetEntity variable

8.   Add a `Finish Execute` node after **Attack Target** and set **Success** to **True**.

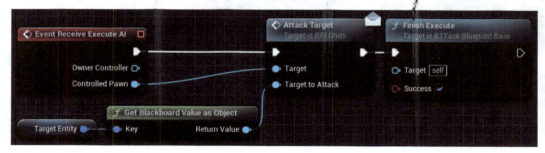

Figure 4.19 – Complete BTT_ShootTarget Blueprint logic

The task is now created. Before we add it to the tree, we need to add a response to the interface call on the character. For now, we will just add a **Print String** node for Shoot:

1. Open the BP_EliteUnit character Blueprint.

2. Double-click **AttackTarget** from the **Interfaces** rollout in **My Blueprint**.

3. Add a Print String node to the new **Event Attack Target** and replace Hello with Shoot.

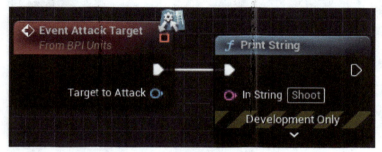

Figure 4.20 – Temporary attack target logic

We will replace this with a proper solution later. For now, this will serve the purpose of confirming that the event is being called by the task in the behavior tree.

Now that we have our task and interface event set up, we can add the task to the behavior tree:

1. Open the BT_EliteUnit behavior tree.

2. Drag from the bottom of the **RootSelector** node and create a new **Sequence** node. Call this node ShootNearbyTargets.

3. Place this new node to the left of the **MoveToTargetLocation** node. This ensures that this node is considered first (i.e., it has a higher priority).

4. Drag from the bottom of the **ShootNearbyTargets** sequence node to add a Rotate to face BB entry task – this will cause the character to face its target.

5. Set the **Blackboard Key** on the **Rotate to face BB** entry task to TargetEntity.

6. Drag from the bottom of the **ShootNearbyTargets** sequence node again and add a BTT_ShootTarget task.

7. Set the **Blackboard Key** on the **BTT_ShootTarget** task to `TargetEntity` if not set automatically.

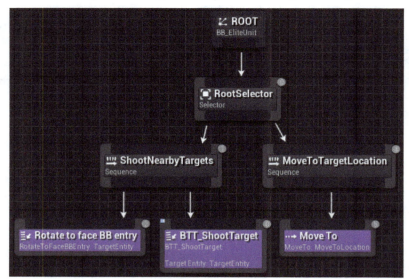

Figure 4.21 – The behavior tree so far with a new sequence and tasks added

The main structure of the branches is now complete; what we don't have, however, is any way of identifying and setting the target identity. We are going to use a service to find our nearest target and a decorator to identify when the `TargetEntity` key has changed in order to choose when to complete the tasks in the `ShootNearbyTargets` sequence.

Identifying enemy targets with a service

To identify the enemies around the character, we are going to place a sphere trace within a service on the **RootSelector** selector node. This will provide the required information to the decision-making processes below as to what part of the tree will run, reducing the need to do tasks that don't return successes.

So, let's start by creating a service:

1. Click the **New Service** button at the top of the **Behavior Tree** editor.

2. Ensure the asset is being saved in the `Blueprint` folder and set **Name** as `BTS_Find NearestTarget`.

3. Open the new asset (if it doesn't open automatically).

4. Create a new float variable called `SearchRadius`, set its default as `500`, and make it **Instance Editable**.

5. Create a new Blackboard key selector variable called `TargetEntity` and make it **Instance Editable**. Just like when creating the *Shoot Target* task, it is vital this is spelled correctly, matching the Blackboard key name.

6. Override the **Receive Tick AI** function by either right-clicking on the graph and searching for the node or using the override dropdown.

7. Add a `Sphere Trace For Objects` node and set it up as shown:

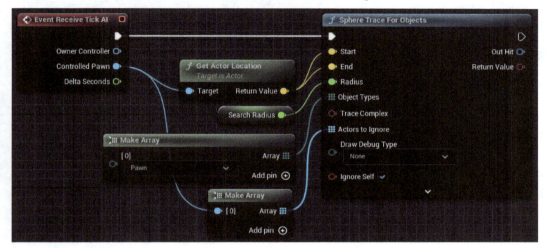

Figure 4.22 – Sphere trace for pawns other than the unit doing the search

8. Check the **Return Value** is **True** and set the **TargetEntity** Blackboard value to the **Hit Actor** output of the **Out Hit** struct pin using a `Break Hit Result` node, as shown:

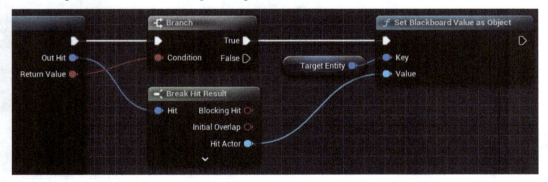

Figure 4.23 – Setting Blackboard key for Target Entity to Hit Actor

9. If the **Return Value** is **False**, clear the `TargetEntity` key using a **Clear Blackboard Value** node. This can be created by dragging from the **TargetEntity** variable node.

The service Blueprint is now complete and looks as shown in the following screenshot. If you would like to visualize the sphere trace in order to determine whether the **Search Radius** variable is suitable, set the **Draw Debug Type** dropdown in the **Sphere Trace For Objects** node to **For One Frame**.

Figure 4.24 – Completed service logic

With the service created, we can now add it to the Root selector:

1. In the BT_EliteUnit behavior tree, right-click the **RootSelector** node, navigate to **Add Service...** and select **BTS Find Nearest Target**.

2. Make sure the **Target Entity** value is set to TargetEntity and that the search radius is showing as 500. The search variable can be changed depending on the needs of the behavior tree it is being added to.

Figure 4.25 – RootSelector with service added

With the service added, the **RootSelector** should now look as shown in the preceding screenshot. You may need to rearrange your nodes to accommodate its increased size.

Now that we have the selector set up to find the nearest target, we can use the result of that service to define whether the shooting part of our behavior tree can be executed.

Adding a decorator to activate the shooting sequence

In order to ensure that the character only tries to rotate and shoot at a target that exists, we need to make sure we only run those tasks when `TargetEntity` has a valid value (this is why we clear the value when the sphere trace returns no hit results). We are going to add a decorator to `ShootNearbyTargets`. A decorator (known as a conditional in other behavior tree systems) defines whether a branch or node can be executed. We will use the decorator to monitor the `TargetEntity` Blackboard key value to see if we have anything to shoot; when we do, the `ShootNearbyTargets` sequence can be executed:

1. Right-click the **ShootNearbyTargets** sequence node, navigate to **Add Decorator...**, and choose **Blackboard**.

2. Select the decorator (the blue box that has now appeared) and rename the node `TargetFound`.

3. Ensure that **Blackboard Key** = **TargetEntity** and **Key Query** = **Is Set**.

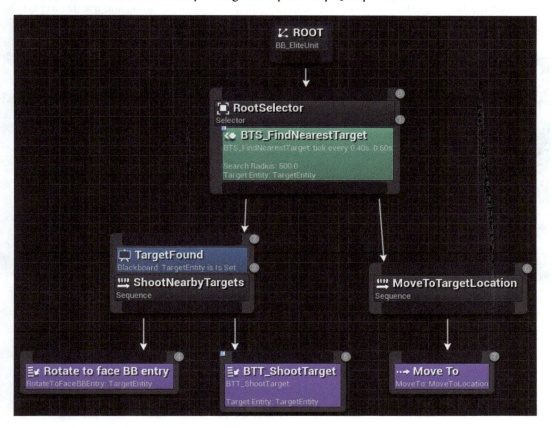

Figure 4.26 – The final behavior tree

If you now test the system by playing in the viewport, you should be able to select and move the green character near either of the red characters. When the green character stops, it will find the nearest target, turn, and should create a print string of `Shoot`. It will continue to do so until the target is no longer valid, which, at this point, will be an eternity as we are not dealing any damage to the unit. You also won't be able to move the character once it finds a target because there is currently no way of getting the AI to move away from the `BTT_ShootTarget` task.

To damage the enemy unit, making use of the `Health` component we created earlier, we need to replace the **Print String** node in the `BP_EliteUnit` **Attack Target** event with some logic to confirm the line of sight and then apply damage:

1. Open the `BP_EliteUnit` character Blueprint.

2. Delete the **Print String** node from the **Event Attack Target** logic chain.

3. Cast the **TargetToAttack** object to **Actor** in order to get the actor's location and use that as part of a **Line Trace By Channel**, as shown in the following screenshot. The line trace has been set up to draw onscreen for 5 seconds so we can confirm the trace is happening correctly.

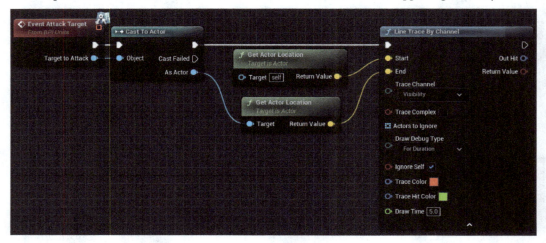

Figure 4.27 – Line Trace By Channel setup from shooting actor to target actor

4. Create a `Break Hit Result` node from the **Out Hit** pin of **the Line Trace By Channel** and compare **Hit Actor** to the **Target to Attack** variable from the **Event Attack Target** node. If **True**, add an `Apply Damage` node with `100` **Base Damage**; if **False**, the trace has hit something else, so at this stage of development, add a `Print String` node for `No Line of Sight`.

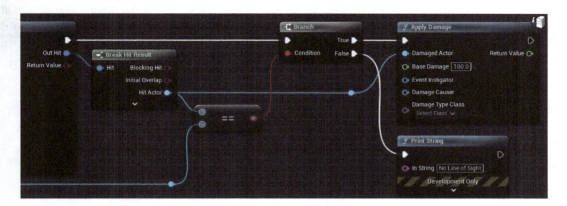

Figure 4.28 – Apply damage to target actor if confirmed. The reroute nodes at the bottom left are routed from the As Actor pin of the Cast to Actor node

5. Lastly, to enable the enemy unit to be destroyed, we need to utilize the **OnComponentDead** event from the **Health** component by linking it to a **Destroy Actor** node:

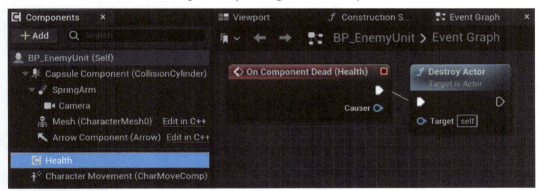

Figure 4.29 – Health component death event linked to a Destroy Actor node

Next, we will look at the final thoughts.

Testing and final thoughts

Testing the game now should remove the enemy instead of continually printing Shoot. This will result in the target entity being cleared, which in turn will allow you to move the character again.

While we now have a successful AI-controlled unit that moves and shoots, there is one other thing that is worth exploring, which is the **Observer aborts** setting on the `TargetFound` decorator. We originally left this set as `None`, which means the unit will complete its **Move To** task before considering shooting. We can change this to `shoot` as soon as it detects an enemy with a simple change of this value:

1. Select the `TargetFound` decorator.
2. In the **Details** panel, set the **Notify Observer** to **On Result Change**.
3. Change the **Observer aborts** to **Lower Priority**.

Test the game again. You should now see that if you tell the unit to move past one of the enemies, it will stop en route to engage with the target. This is because the decorator has identified a change in the value of `TargetEntity` from the `BTS_FindNearestTarget` service and can now interrupt any lower-priority branch of the behavior tree, so any node which is to the right of it in the selector, which in this case includes the `MoveToTargetLocation` sequence, which holds the **Move To** task.

Summary

In this chapter, we have discussed the game development patterns that are present as premade tools and systems inside Unreal Engine. Namely, the component, update, and behavior tree patterns.

We made a `Health` component that will hopefully prove useful in your future projects, along with a simple behavior tree showing each major piece of the system. Armed with this knowledge, you should be able to make a functional AI system for any of your future Unreal projects.

The next chapter will take what we have learned about the update method and explain how we can do better. What is the impact of using `Tick` within the engine and how can you measure that impact?

Part 2: Anonymous Modular Design

In this part, we will begin creating more efficient code to build modular solutions that communicate with each other efficiently and cleanly, without the use of direct references and casts.

We will start by reducing the need for gated polling in our code by replacing it with cleaner communication methods, and we will explore interfaces and event delegates to allow our code to be decoupled. This will culminate in a coding design paradigm we call Anonymous Modular Design, which should make our object-oriented code the most maintainable and expandable it can be.

This part has the following chapters:

- *Chapter 5, Forgetting Tick*
- *Chapter 6, Clean Communication – Interface and Event Observer Patterns*
- *Chapter 7, A Perfectly Decoupled System*

5

Forgetting Tick

In this chapter we are going to explore the topic of tick (an event or process that is called every frame), why using it can cause frame-rate issues for our game projects, and two approaches to building systems without using tick, reducing the impact of systems on the CPU. We will consider the example of a guard tower in an RTS game, with a searchlight rotating back and forth. If the player's unit is seen by the searchlight, it will stop its rotation. We will start with the Update pattern introduced in the previous chapter and iteratively improve from there, reducing our reliance on the Update pattern by using Timers, Timelines, and event-driven solutions instead. At each step, we will measure code efficiency, using execution counting to quantify our improvements.

In this chapter, we'll cover the following topics:

- Tick events and the challenges we face when using them
- Timers and Timelines
- Event-driven systems

Technical requirements

In this chapter, we will focus on the Guard Tower classes in the *Chapter 5* folder of the project, linked via GitHub, that we downloaded for previous chapters. If you haven't followed along with the developments in the previous chapters, you can download a version of the project, ready to start the *Chapter 5* tutorials from the chapter5 branch on GitHub at https://github.com/PacktPublishing/Game-Development-Patterns-with-Unreal-Engine-5/tree/main/Chapter05

Before we begin

In this class, we will use a new smart pointer type. `TObjectPtr<>` is the Unreal Engine 5 standard way of dealing with components held on an actor. The prior standard was to use raw pointers, but this will no longer work with the new garbage collector, as the new system provides reference tracking to detect when an object is actually used. This can mean that objects only stored in raw pointers get mistaken for de-referenced memory and can be deleted while still in use.

Within functions, we will still use raw pointers, as these variables will go out of scope, and so the garbage collector is not necessary to free their memory. Member variable object references will be stored as `TObjectPtr<>` and asset references stored as `TSoftObjectPtr<>`.

A world that ticks

In the last chapter, we covered the idea of the Update pattern. When using the Update pattern, we design with the philosophy that lots of different types of classes need to execute code on each frame. To achieve this, we will use an update function or, in the case of Unreal, a **Tick Event**. This pattern provides a fast way to make things happen, which is why it has a certain gravity, pulling many learning resources into leaning on it heavily. The volume of learning materials doesn't make this a good idea, but it does allow systems to be propped up quickly, thus making learning from resources like this easy to understand. The problem lies in the seeming loss of understanding, over the cost of relying on this method. When we place any nodes or lines of code under the purview of tick, we need to remember it runs once per frame. On modern machines this equates to an average of 60 times a second, but it can vary depending on hardware, which is an important consideration for developers. For small games on modern hardware, the odd variable set or transform change may be fine; however, this is significantly more concerning if you consider the impact of a loop. This should make it clear that, as powerful as tick may seem, a web of problems lies beneath its surface.

To illustrate the problem, let's look at the `AGuardTower_CH5_1` actor in the *Chapter 5* folder. We have followed best practice with the header file, using forward declared classes for member component definitions and the most limited property specifiers that we can. You will probably notice though that the only function, bar the constructor, is `Tick`. In this function, we carry out all our searchlight functionality. Stepping through the tick function on the searchlight, we have the following:

- A call to the parent Tick
- A sphere trace that gets reference to any actor that falls within the rough area of the light
- A check to see whether the detected actor is of the player pawn type
- Code branches to rotate the searchlight in every direction when the player has not been found

The GuardTower body file contains a tick function which looks like this:

GuardTower_CH5_1.cpp Tick function

```cpp
void AGuardTower_CH5_1::Tick(float DeltaTime)
{
    Super::Tick(DeltaTime);

    FVector startLocation = _Arrow->GetComponentLocation();
    FVector endLocation = _Arrow->GetComponentLocation() +
        (_Arrow->GetForwardVector() * _DetectionRange);

    FHitResult hit;
    UKismetSystemLibrary::SphereTraceSingle(GetWorld(),startLocation,
    endLocation,_DetectionRadius, UEngineTypes::
        ConvertToTraceType(ECC_Visibility), false, {},
            EDrawDebugTrace::ForOneFrame,hit, true);

    ACharacter* otherCasted = Cast<AEliteUnit>(hit.GetActor());
    _EnemySpotted = (otherCasted != nullptr);

    if (!_EnemySpotted)
    {
        if (_RotateForward)
        {
            _LightPivot->AddLocalRotation(FRotator(0.0, 0.2, 0.0));

            if (FMath::IsNearlyEqual
                (_LightPivot->GetRelativeRotation().Yaw, 40.f))
            {
                _RotateForward = false;
            }
        }
        else
        {
            _LightPivot->AddLocalRotation(FRotator(0.0, -0.2, 0.0));

            if (FMath::IsNearlyEqual
                (_LightPivot->GetRelativeRotation().Yaw, -40.f))
            {
                _RotateForward = true;
            }
        }
    }
}
```

This code does run, and you can see that if you drag an instance of `BP_GuardTower` into the level provided in the *Chapter 5* folder. However, there are some issues, so let's break down these problems in order:

- With all the logic being on tick, we are performing some costly actions often. A sphere trace on its own is not too costly, but if we perform that sphere trace once per frame, per guard tower in a scene, the computational cost can add up quickly.

- Quite a few getter functions are repeatedly used without caching. The world pointer could have been cached on `BeginPlay`, as that is unlikely to change, and multiple calls to the arrow component's location could be done once. These are only minor improvements.

- Casting is nowhere near as costly as it used to be in the early days of Unreal Engine. That said, when casting to a class, the target class must be loaded to ensure a match. This must be kept in mind when casting to a large class, as it can inflate the size of the process in RAM. Alternatives could include casting to an interface if you only need some functionality, checking tags if it is just a validity check, or, even better, trying to flip the communication on its head and remove the need to cast.

- We then have a pattern that we will refer to as **gated polling**. On tick, check to see whether a block of code needs to run; if so, run it. This gated polling pattern is repeated a couple of times for different blocks of code. The alarm bells should now be ringing but maybe for the wrong reason. Yes, we have a repeated pattern of code, and we previously talked about avoiding repetition; that isn't the real issue here. Any time we process something that may or may not need to happen, we guarantee a wasted check on some frames when it fails. We should consider how we can avoid the check and only process the code when it does need to happen.

- Lastly, we have a compound problem with each branch of the gated poll. Yes, we are hardcoding values, both the rotation limit and the speed of rotation. Turning the hardcoded `0.2` and `40` into float member variables called `_RotationSpeed` and `_RotationLimit` respectively is a simple fix for the first part. Adding `EditAnywhere` to the `UPROPERTY()` block above each new variable will allow designers to not only balance the values in the actor but also create varied instances in their levels. This still leaves an issue with our code. Tick is not constant, yet we are using a constant value for rotation. This is where `DeltaTime` comes in. It is an argument of tick passed in as the time since the last frame rendered. Multiplying values by this will cause them to be applied evenly across 1 real-world second.

It should be pretty clear that there is a lot of room for improvement with this code, but how much room? Back in *Chapter 3*, we discussed Big O notation, but to get a higher resolution analysis, we need to calculate the $T(n)$ or time efficiency of the algorithm. This should provide us with a baseline to prove that we can quantifiably do better.

The first sphere trace section contains roughly three assignments, six function calls, three arithmetic operations, and the internal time of a sphere trace. The cast section is two function calls, one test, and one assignment. In the `if` block, both branches are basically the same number of executions, so we will only count the top branch, giving us two tests, four function calls, and one assignment. Altogether, this makes roughly 23 + 1 * sphere trace executions per frame, with no overhead beyond a standard setup. Over a second at 60 **frames per second** (**FPS**), we would be running 1380 + 60 * sphere trace executions.

> **Important note**
> You can find the full working solutions for the following two sections under the _2 and _3 versions of the GuardTower_CH5 class, but try to follow along with this section, making changes to the _1 version to get the most out of the chapter.

Now that we've reviewed the problems with using Tick for the guard tower, let's begin to look at how we can fix it, removing our reliance on Tick and writing better code.

A clock that stops

Now, we'll deal with this gated polling issue. The go-to solutions for this are **Timers and Timelines**. Timers provide a way to delay the calling of a function, but not in the same sense as the **Delay** node from Blueprint. The C++ backend for **Delay** nodes can be found in UKismetSystemLibrary, where you will see that simply calling it is a bit of a hassle. Timers will perform a check on each frame they are active to see whether the function they point at should fire yet. This can be a useful behavior for dynamically set delays or systems where you only want a signal after a set amount of time, such as a countdown to decrease once per second instead of once per frame. Timelines, however, provide a way of processing a behavior, similar to an update while a curve is being queried. The length of these curves is predetermined, although the play speed of the Timeline can be altered to achieve a dynamic length. Timelines can also hold multiple types of tracks, which will all be synchronized when playing, allowing a single Timeline to drive a lot of elements. This provides a better fit for our problem, as we need to retain the update behavior but with a smaller footprint, reducing the number of checks we perform each frame.

Looking at the following code, we can see that to set up a Timeline in C++, we start by adding the following variables and functions to our header to facilitate the Timeline. It may seem at first glance that adding this much to a class would slow the process down, but each element has its part to play. The first two variables are delegates and will provide us with a way of linking functions to the Timeline dynamically. These are followed by the functions we will link. This pattern follows a naming convention described as follows to aid future readability:

- On<object name><delegate purpose> for the delegate
- Handle_<object name><delegate purpose> for the linked function

The following two functions, StartRotation and StopRotation, are simply there to act as wrappers for the Timeline methods.

The Timeline itself is created as `TObjectPtr<>`, allowing for forward-looking reference tallying and safer garbage collection. The property specifiers on this also don't matter too much, as we access all the functionality we need through functions we have created. The last variable is `TSoftObjectPtr<>` to `UCurveFloat`. This type is used by the engine to store a spline in 2D space as a set of keyframes with tangents. We will use it to drive our searchlight angle over time, but we want a designer to have access to make changes. This means we need to store the variable as a reference to an asset that does not yet exist, hence the `TSoftObjectPtr<>` wrapper. Setting this as `EditAnywhere` will allow the default value and the instance variants to be set via their respective editor panels:

GuardTower Timeline Header

```
private:
    FOnTimelineFloat onTimeline_Update;
    FOnTimelineEventStatic onTimeline_Finished;

    UFUNCTION()
    void Handle_RotateLight_Update(float val);

    UFUNCTION()
    void Handle_RotateLight_Finished();

    void StartRotation();
    void StopRotation();

protected:
    UPROPERTY()
    TObjectPtr<UTimelineComponent> T_RotateLight;

    UPROPERTY(EditAnywhere)
    TSoftObjectPointer<UCurveFloat> _Curve;
```

Now, we will turn our attention to the `GuardTower` constructor implementation shown here. We need to create the Timeline component in the constructor. Then, our delegate handles need to be bound to the listener functions. Binding like this will allow rebinding later down the line as necessary, while any class derived from this will always have a default response. It is also worth noting that binding is done via function names created as `FName` types. This will make spelling and capitalization important, and it is unlikely that your IDE will have auto-complete functionality for these:

GuardTower Timeline Constructor

```
T_RotateLight = CreateDefaultSubobject<UTimelineComponent> (TEXT("T_
RotateLight"));
onTimeline_Update.BindUFunction(this, FName("Handle_RotateLight_
```

```
Update"));
onTimeline_Finished.BindUFunction(this, FName("Handle_RotateLight_
Finished"));
```

We cannot do any more setup inside the constructor, as we are blocked by the fact that we could not have set a value for the `_Curve` variable yet. This will require linking in the editor, which forces the rest of the initialization process into the `BeginPlay` method.

The first thing to consider is that the variable may not have had a value set at all, which would invalidate the need to continue any setup. Instead of nesting all the code inside a positive conditioned `if` statement, we will reverse the condition to check for a negative condition and plug in an early return. This makes our code more human-readable and reduces the amount of information we need to keep in mind while constructing the function.

Following that, the Timeline needs to have a few things set up, as shown in the following code block. Any tracks that it should be running need to be added. Here, we will add a float track based on whatever asset the `_Curve` variable references. This needs to be injected with the delegate it will call whenever its value changes and the name of the track for future reference. Here, we will plug in the update delegate, `onTimeline_Update`, and a new `FName` value that makes sense for our usage. The same then needs to be done for the timeline's finished callback. It is slightly different from the other function in that it doesn't require a curve or a name, just a callback. So, all we do here is insert the finished delegate, `onTimeline_Finished`.

In this case, we also need to set the looping nature to `false` and the ignore time dilation property to `true`. We will control how the Timeline replays, so there is no need for any auto behavior, and the other variable is mainly there to show what sort of control you can get with a timeline. There are plenty of other properties to fiddle with to fine-tune how timelines behave, so it is well worth your time exploring the type header for functions.

> **Useful tip**
>
> If you want to explore type or class headers, define a variable of that type anywhere in your code. This will allow you to right-click the type and select **Go To Definition** (*F12* in Visual Studio and Rider).

All these changes should lead you to a `Beginplay` setup for the Timeline that looks like the following code:

GuardTower Timeline Setup

```
Super::BeginPlay();

if (_Curve == nullptr)
{
```

```
    return;
}

T_RotateLight->AddInterpFloat(_Curve, onTimeline_Update,
    FName("Alpha"));
T_RotateLight->SetTimelineFinishedFunc(onTimeline_Finished);
T_RotateLight->SetLooping(false);
T_RotateLight->SetIgnoreTimeDilation(true);

StartRotation();
```

Now, we need to fill out the callback handler functions, which are quite simple in nature – one to rotate the light on its killable update and the other to change the light's direction when it reaches the rotation limit:

Callback handler functions

```
void AGuardTower_CH5_3::Handle_RotateLight_Update(float val)
{
    _LightPivot->SetRelativeRotation(
    FRotator(0.f, FMath::Lerp(-40.f, 40.f, val), 0.f));
}

void AGuardTower_CH5_3::Handle_RotateLight_Finished()
{
    _RotateForward = !_RotateForward;
    StartRotation();
}
```

The last thing we need to do is call the `StartRotation()` function to start the Timeline. The body of this simply calls the timeline's `Play()` function if `_RotateForward` is `true`; otherwise, it calls `Reverse()`. We also have a wrapper for the `Stop()` function, but this provides no benefit beyond being good practice for later development, all it does is call the `Stop()` function on the timeline. With that done, we can delete the nested `if` statements at the end of our old `tick` function, add a `StopRotating()` call to our sphere trace where it sets `_EnemySpotted` to `true`, and head to the editor to link up our `_Curve variable` with an asset.

With the Timeline working, we can leverage some of its other features to improve the gameplay as well. The old rotation behavior on tick was very linear and not good for gameplay. Controlling the rotation with an f-curve telegraphs future actions to the player. As the rotation slows, the player gets an innate understanding that the light is about to swap direction and can plan accordingly. This concept of **telegraphing**, although it is a game design term, is important for programmers and system designers to understand, as it will aid us in building tools that are more useful to designers. Giving designers access to this curve will allow them to drastically change behavior with little effort, making better gameplay overall. The curve asset we have created in the project for you is called `C_GuardTower_RotAlpha` and can be found with the guard tower class in the *Chapter 5* folder. The curve is a simple f-curve with flat tangents on start and end points, covering (0, 0) to (5, 1), as shown in *Figure 5.1*.

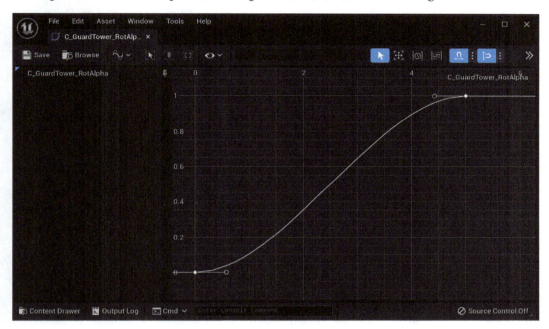

Figure 5.1 – The C_SearchLightRotation curve asset in the curve editor window

Now that we have changed the old, gated polling out for a Timeline system, we should recalculate the *T(n)* performance. We removed two tests, four function calls, and one assignment, bringing the total down to 16 + 1 * sphere trace. Then, we added a timeline, which is worth one test and one function call per frame, one test and six function calls that will run one time, one test, and two function calls that will currently run once every 5 seconds. Simply adding the values back together makes the situation look worse, giving us a 28 + 1 * sphere trace, but this is wrong. The new code will not run as regularly as the old, and when multiplying it out by the correct frequencies, you get (60 * (18 + 1 * sphere trace)) + (3 / 5) = ~1081 + 60 * sphere trace per second, plus 7 overhead. This is roughly a 22% improvement on the first implementation.

This is good, but we still have a problem. There is still something we are doing every tick: the sphere trace. Every frame this projects a sphere through space to see whether the player's character is in "view" of the tower. This setup mirrors our previous problem of gated polling. Every frame, we run the test. If the player is overlapped, we fire the seen logic. If the player is not detected or is just out of sight, then nothing happens, and that sphere trace we checked was wasted.

The following section will go through the process of improving this and the architectural decisions that govern which route to take when considering efficiency versus design.

Event driven systems

A software engineer's job is to solve problems. Most of the time, this means special treatment in certain cases to make sure eventualities are handled. Sometimes, perfection is found, and the problem is not solved but erased from existence. In our current system, we want to solve the problem of needing to trace a sphere through our world for the player in every frame.

The question is not, *how do we make this more efficient?* Instead, we should be asking, *why sphere trace?* There isn't really a clear answer to this. Yes, it allows us to check the volume of space for a player, but so do colliders. Yes, we can check all the space between the light mesh and the floor, but in our game, the player cannot jump. Yes, we can tell whether there are objects blocking a view of the player, but so can a cheaper line trace on the frames when we are not sure.

From these three answers, we can form a new solution. In principle, we will attach a sphere collider, as it's the cheapest primitive to use, to the light. As the light swings, so does the attached collider. When the collider overlaps the player, we use a line trace from the light to the player to see whether there is a clear line of sight. This effectively flips the interaction between the searchlight and the player on its head. No longer does the searchlight ask the world whether it can see the player every frame. Now, the world will tell the searchlight only when the player is seen.

Walking through our implementation, we will link a handler function to our new sphere component's `onComponentBeginOverlap` event. This handler function, detailed in the following code, starts with an early exit in case we have already collided with a player unit. If so, then we can ignore future collisions. In a larger scenario with more colliders in the scene, you may want to ignore this check and move on to incrementing a counter instead, as the searchlight is big enough to cover multiple units at once. We then cast the overlapped actor to the class we know the player will use. This, again, could be switched to a check for `teamID` or some similar group-identifying value. This would allow multiple players or factions to use the same units. The next few lines set up the values for and execute a line trace through the world, from the light position to the center of the unit we have just overlapped with. This is done because the collider we will swing around has no idea of the map it is on and where any vision-blocking walls may be. We will use the line trace as another early return because if it does hit something, that means we can ignore the player collision due to a mesh blocking vision. With a clear line of sight established, we then simply set the `_EnemySpotted` variable to `true` and stop the timeline-driven rotation. This would be where you would link into something like an alert system to draw the attention of nearby friendly units.

Let's start with the Begin Overlap:

GuardTower Sphere Begin Overlap function

```cpp
if (_EnemyUnit != nullptr)
{
    return;
}


_EnemyUnit = Cast<AEliteUnit>(OtherActor);

if (_EnemyUnit == nullptr)
{
    return;
}


FHitResult hit(ForceInit);
FVector start = _Arrow->GetComponentLocation();
FVector end = _EnemyUnit->GetActorLocation();

if (UKismetSystemLibrary::LineTraceSingle(
        GetWorld(), start, end,
        UEngineTypes::ConvertToTraceType(ECC_Visibility),
        false, {_EnemyUnit}, EDrawDebugTrace::ForDuration, hit,
        true, FLinearColor::Red, FLinearColor::Green, 0.5f))
{
    return;
}


_EnemySpotted = true;
StopRotation();
```

The onComponentEndOverlap equivalent event is handled by the function shown in the following code block and is much smaller, as it only needs to check that the actor leaving the collision is the one that triggered the alarm in the first place. Then, it can reset the _EnemySpotted and _EnemeyUnit variables, followed by continuing the Timeline rotation code:

GuardTower Sphere Overlap End

```cpp
if(_EnemyUnit != OtherActor) {
    return;
}
    _EnemySpotted = false;
```

```
_EnemyUnit = nullptr;
StartRotation();
```

> **Important note**
>
> Component overlap functions were explored at the end of *Chapter 1*. Remember to bind only functions matching `FComponentBeginOverlapSignature` and `FComponentEndOverlapSignature`, respectively. The arguments used in the sample code follow this pattern: `UPrimitiveComponent* OverlappedComponent, AActor* OtherActor, UPrimitiveComponent* OtherComp, int32 OtherBodyIndex, bool FromSweep, const FHitResult& SweepResult` for the begin overlap, which we can see are the same, minus the last two arguments for the end overlap.

This last change, from traces to colliders, calls for another recalculation of the $T(n)$ performance. We repositioned the rest of the functionality in the Tick function completely, leaving only the Timeline update running. This drops our per frame executions to 2! The processing hasn't been completely removed though. In the worst-case scenario, the player would overlap with the sphere collider and leave the interaction within a second, meaning all the code could fire. This would be equivalent to four assignments, eight function calls, three tests, and 1 * line trace on begin overlap and two assignments, one function call, and one test on end overlap. This, plus the four function calls to set up the new component and bind these functions, detailed in the following code, brings our cost to 2 per frame, 19 + 1 * line trace per second, and 3 per 5 seconds. This makes (60 * 2) + (19 + 1 * line trace) + (3 / 5) = ~140 + 1 * line trace per second, with an overhead setup cost of 11. This is a massive 90% improvement on the first implementation, showing how a change in approach can make a massive impact on performance, even though it may seem like more has been created. For reference, the following code can be used to link handler functions to the relevant component overlap events if you need a syntax example. Your functions may have different names, but these lines must run for the callback to work:

GuardTower constructor new lines for sphere component setup

```
_Sphere = CreateDefaultSubObject<USphereComponent>(TEXT("Sphere"));
_Sphere->SetupAttachment(_LightMesh);
_Sphere->OnComponentBeginOverlap.AddDynamic(this, &AguardTower_
CH5_3::OnSphereOverlapBegin);
_Sphere->OnComponentEndOverlap.AddDynamic(this, &AguardTower_
CH5_3::OnSphereOverlapEnd);
```

The performance benefit here is shown to be quite extreme, but this is actually the norm when compared to the prototype approach of dealing with all behavior in Tick. We can also apply this technique to any situation where the Update pattern has been extensively used. Unreal Engine 5 already uses it within the Enhanced Input system, only firing the delegate callback events when the input has been pressed. The old polling system would fire on axis inputs regardless, leaving us to gate values that were not needed. We can also look at a few places where Unreal provides options for communication, namely UI and networking – two very different areas, but the principle is the same. Both provide a way of quickly setting up a link using a polling, or update, method that fires every frame. For UI, it is property binding, and for networking, we have replicated variables. In both cases, there is some extra function run every frame to see whether the variable it is linked to has updated. Hopefully, the parallels are clear with the example we have just explored. Specific solutions would be to manually update UI through a custom function call and use replicated functions when dealing with networked systems, also called **Remote Procedural Calls** (**RPCs**), which allow us to only send data across a network when it needs to be updated.

With this chapter complete, you should now understand the need, and method, to design systems with an event-driven approach. We have shown that we can make code that has the exact same functionality in many ways, but using an event-driven approach can save a lot of processing, even if it means a more complex setup. As an additional task, consider how to make the searchlight work for multiple controllable units, or how it could track a unit once it has been seen until it loses sight. On a higher level, how could the searchlight notify the surrounding enemies that it has seen your unit? It is advised that you revisit this class after each chapter to see how the tools learned later can improve this further.

Summary

Through the explicit example of a watchtower in our example game, we learned a host of different techniques and considerations to improve efficiency in our systems. We applied the general technique of always trying to do less, by swapping a gated polling system for a killable update and event-driven systems. We used Unreal Engine-specific tools, such as the Timeline and Timers, to achieve these patterns, and we thought about how their application may affect the design of the gameplay beyond our numbers-based efficiency targets. The iterative process of improvement that this chapter focused on has also shown how to improve code in small measurable ways, using $T(n)$ calculations to quantify each step. Moving forward, we will look at more general programming tools and their implementations within Unreal to expand this toolkit. The big takeaway is to break the problem down into its base parts and measure your improvements as you go. This will allow you to better communicate your process to, and work with, your team to achieve your goals.

6

Clean Communication – Interface and Event Observer Patterns

In this chapter, we will explore two design patterns that allow us to improve how actors communicate with each other. Utilizing the interface and event observer patterns to allow us to build better, cleaner communication solutions, reducing memory and processing cost of communication. To explore these, we will expand on the RTS game by implementing a health component that can be added to any actor. We will start by looking at interfaces in C++, following the concepts of the interface design pattern, and then explore the event observer pattern, with an exploration of Event Dispatchers in Blueprint and the implementation of event delegates in C++.

In this chapter, we will cover the following topics:

- Interfacing communication across classes in UE5

- Implementing event delegate communication across UE5

Technical requirements

In this chapter, we will bring together the isolated systems we have built in previous chapters. Presuming you have followed along, you should be able to tie it all together in this chapter.

If you want to jump straight in at this point, you can download the `chapter6` branch from GitHub at `https://github.com/PacktPublishing/Game-Development-Patterns-with-Unreal-Engine-5/tree/main/Chapter06`

Interfacing communication across classes in UE5

Interfaces are a class that facilitates communication between different classes, without either class being aware of the other's type. An interface class holds a set of common functions that can be implemented on any class through polymorphism.

To abstract this, let's think about a ridiculous situation where you have a stack of pages, and you are asked to verify that it is indeed a copy of *War and Peace* that is just missing the cover. The only way to be truly sure is to compare each word on every page with a verified copy. This is time-consuming and along the lines of what Unreal Engine does when casting a class to another type.

> **Important note**
> This isn't exactly what the engine does. There are efficiencies built in that speed this process up, ensuring that it is not the entire class that is checked, but the point still remains that casting can be needless.

If we dig deeper into this hypothetical, knowing why you were asked to verify the book's nature can make a big difference. If it was to verify the book's integrity, then there is no way around it, but if the request was just to see whether a certain chapter existed, then we don't need to do all that much work. Instead, we just need to look at the contents page, skip to the requested chapter, and read from there. This is what an interface does; it's like the contents page of a book. The book promises that the required chapter will be in the area signposted by the contents page.

One of the most-used solutions to build cleaner methods of communication between two actors is to implement Blueprint interfaces, which remove the need to cast to an actor's specific class, allowing you to use an actor reference instead. Let's take a deeper look at Blueprint interfaces before we move on to interfaces in C++.

Blueprint interfaces

We've previously made use of a **Blueprint interface** in *Chapter 2* while improving the cascading cast chain problem, at which point we created a Blueprint interface, set up a function definition, and used it as a callable event on the various weapon Blueprints. The event was then called from the character Blueprint, allowing it to call the event on an attached weapon (a child actor component) without needing to cast the reference to a specific class, improving the efficiency of the system as well as reducing the memory impact.

So, from *Chapter 2*, we already know how to create a Blueprint interface, create functions within the interface, and call them as events, but let's look at a few of the features of a Blueprint interface that we haven't encountered yet.

Let's explore this using a new interface. Go ahead and create a new Blueprint interface asset, and call it BI_Test, as this is only a temporary class for exploration purposes.

Before we add any functions, let's take a look at the various settings by clicking on **Class Settings** from the top menu, which will change the details panel to show a range of variables associated with the Blueprint class.

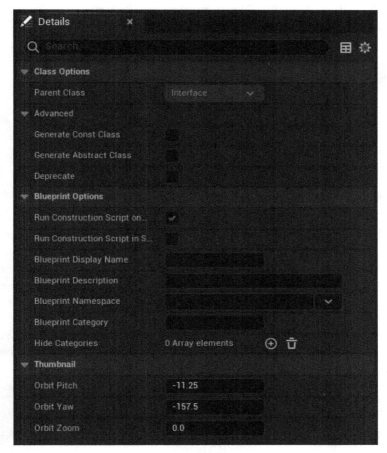

Figure 6.1 – The Blueprint interface Class Settings showing in the Details panel

The two most beneficial settings when building systems are **Blueprint Display Name** and **Blueprint Description**, as both provide improvements to the user experience when working with Blueprint Interfaces. Without these being set, Unreal will display only the asset name and provide no description, which, while okay for a solo developer in the moment, is less helpful later in development or when working with other people.

In order to see the settings working for yourself, we suggest creating and using a test Blueprint class, `BP_Test`. This can then be deleted once you are confident you understand the settings.

Let's take a look at how we set these up and what effect they have on the user experience.

Blueprint Display Name

The **Blueprint Display Name** setting provides a string input to allow you to define how the interface appears in the **Class Settings** of a Blueprint class when the interface has been implemented.

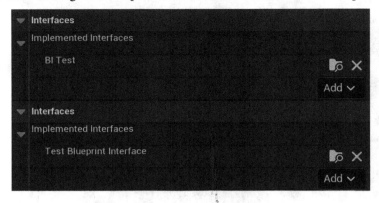

Figure 6.2 – Class Settings of a Blueprint class with an implement interface with and without a Blueprint Display Name

Figure 6.2 shows the comparison between the class settings of a Blueprint class for the end user when implementing an interface with or without a **Blueprint Display Name**. Here, the Blueprint interface class `BI_Test`, in the left image, does not have a **Blueprint Display Name**. For the right image, the **Blueprint Display Name** has been set to **Test Blueprint Interface** which is then displayed in the `implemented interfaces` list when the `BI_Test Blueprint` interface has been selected from the **Add** dropdown.

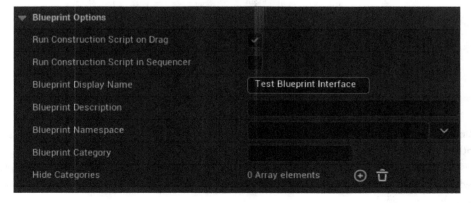

Figure 6.3 – The Blueprint Options rollout of BI_Test with the Blueprint Display Name set.

This allows us, as systems designers, to maintain core naming conventions whilst ensuring that it is always clear what interface has been used in a format suitable for designers and other team members.

Blueprint Description

The **Blueprint Description** setting provides a string input, allowing you to define the mouseover hint when choosing an interface to implement, from the **Add** drop-down menu in the class settings of a Blueprint class.

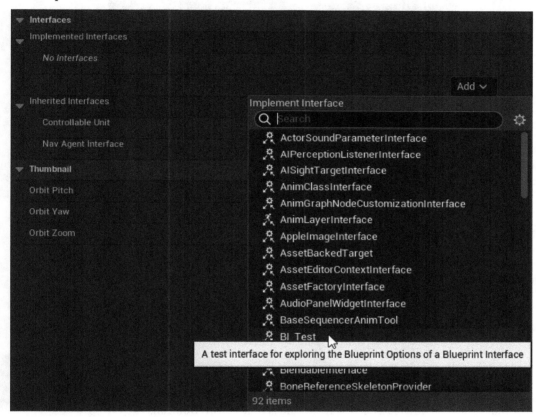

Figure 6.4 – The class settings' Add dropdown when selecting an
interface to implement, with a Blueprint description

Figure 6.4 shows a comparison between the **Add** dropdown for the end user in **Class Settings** when selecting an interface to implement with and without a Blueprint description. Here, the Blueprint interface class, BI_Test, on the left-hand side of *Figure 6.4*, does not have a Blueprint description. On the right-hand side, the Blueprint description has been set to **A test interface for exploring the Blueprint Options of a Blueprint interface**, which is then displayed when the mouse is held over BI_Test in the **Add** dropdown.

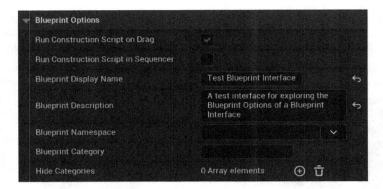

Figure 6.5 – The Blueprint Options rollout of BI_Test with the Blueprint Description set

This allows us to make it clear, when selecting an interface, what it should be used for. The messages can include specifics such as `Required on all classes required to interact with doors` or more general messages, such as `Adds functions for shooting weapons`. The key here is to make sure the description helps other users identify what the interface provides.

Interface events versus functions

When adding functions to a Blueprint interface *functions* list, it can result in either a function or an event when implemented into a Blueprint class.

Using the `BI_Test` and `BP_Test` classes, we can explore the differences in a safe space. Let's begin:

1. Let's start by adding two functions using the **+Add** button; name these `ExampleEvent` and `ExampleFunction`.

2. Select `ExampleFunction` from the *functions* list, and add an output variable to the **Outputs** list by clicking the + icon.

3. Name the new variable `BoolOutput`, and set the type to **Boolean**.

This will have created both an event, and a function for use in a Blueprint. Once you add an output variable, the function will behave as a Blueprint function, and without any outputs, it will act as an event.

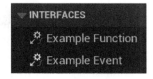

Figure 6.6 – An interface function and interface event in a Blueprint class

Note how the icon is different for a function (a gray icon) and an event (a yellow icon). This isn't the only difference; adding logic to a function works differently to an event. Double-clicking on the event will result in an event node being added to the **Event Graph**:

Figure 6.7 – An interface event when added to the Event Graph

Double-clicking on the function will result in a new tab opening with the function graph, including a **Return Node**.

Figure 6.8 – An interface function with Return Node

When calling an interface function or event, they will appear slightly different in the Blueprint graph. The main differences are the icon and the inclusion of output exec pins (the white outlined arrows on the right-hand side of the node):

Figure 6.9 – A comparison of interface event and interface Function nodes in a Blueprint graph

There are two other possible visual results for Blueprint nodes from interface functions; these are compact nodes that occur when a Blueprint interface function has a **Compact Node Title**.

To test these out, create two new functions in the `BI_Test` Blueprint interface; call these `ExampleCompactedEvent` and `ExampleCompactedFunction`, adding a Boolean variable to the output list as before. This time, however, set the Compact Node title of each function. For now, duplicate the function names with spaces between the words.

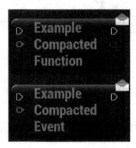

Figure 6.10 – Compacted node examples of an interface event and interface function on a Blueprint graph

Compact nodes, when placed, appear as shown in *Figure 6.10*. Compact nodes don't have titles or pin names; these are not commonly used when working in Blueprint but are available if desired.

Interfaces in C++

In C++, things get a little weird. The original core of Unreal is closer to Objective C than C++, and it shows the most when dealing with interfaces. It is advised that interfaces should be made from the editor, as there is so little that we need to add to the file from the template that typing it all out from scratch is a waste of time.

In the editor, open the section of your content for C++ classes. Right-click and add a new C++ class. When selecting what parent to inherit from, scroll to the bottom of the common classes and choose the **Unreal Interface**. This will create a new header file, like the following template:

Unreal interface template project specific elements replaced with tokens

```cpp
#include "CoreMinimal.h"
#include "UObject/Interface.h"
#include "<class name>.generated.h"

UINTERFACE(MinimalAPI)
class U<class name> : public UInterface
{
    GENERATED_BODY()
};
```

```
class <module name>_API I<class name>
{
    GENERATED_BODY()
public:


}
```

As you can see in the template, there are two classes in the same header file. The top class is created with your chosen class name, prefixed with a *U*. This inherits from the `UInterface` class and allows your class to become part of the Unreal interface system, working with functions such as `UKismetSystemLibrary::DoesImplementInterface`. The second class extends your module and has a similar name, except it starts with an *I* instead. This is the class you should add public functions to and multi-inherit from.

When calling a function from an interface, it is assumed that the reference held is not for the correct type of class and, instead, is most likely an `AActor*`. If this is not the case, then the interface provides no benefit, and the communication strategy should be rethought. Unreal provides a few ways of calling functions via interfaces. The two main methods are casting to the interface as a type and calling the instance of the function, or statically calling the function. Casting to the interface type is exactly as it sounds and would involve something like this:

```
if(IInterface* cachedRef = Cast<IInterface>(actorRef))
{ cachedRef->Function(arg1); }
```

Note how the `Cast` returns the casted pointer and a boolean value for the `if` statement. This works because an `if` on a pointer in Unreal will automatically check whether the pointer is not `nullptr`. However, it does still use a cast, no matter how small it may be. It is faster to use a static function call such as the following:

```
IInterface::Execute_Function(actorRef, arg1);
```

As you can see, this gets to the function call much faster and with less temporary variables declared. The trade-off is the security of the call. In this method, we assume that `actorRef` implements the interface that we call the function from. This assumption means that if the class in question doesn't implement the needed interface we will get a logic error. The solution is to either design in a way that avoids this situation or to check for the interface without casting to it, using the aforementioned `DoesImplementInterface` function in `UKismetSystemLibrary`.

So, with these general concepts in mind, let's look at creating our own interface in C++.

Building an example interface communication

To implement something like the preceding example where actors can be flammable, we can start with an interface called `IFlammable` that will need a public virtual function called `Ignite`. This would look like the following:

Iflammable.h

```
#include "CoreMinimal.h"
#include "UObject/Interface.h"
#include "flammable.generated.h"

UINTERFACE(MinimalAPI)
class UFlammable : public UInterface
{
    GENERATED_BODY()
};


class EXAMPLE_API IFlammable
{
    GENERATED_BODY()
public:
    virtual void Ignite();
}
```

To implement this in other classes, you simply need to use polymorphism and override the interface functions. The following example shows a tree actor multi-inheriting from the interface to implement it:

Excerpt from possible example flammable interface usage Tree.h

```
#include "flammable.h"
UCLASS()
class Example_API ATree : public AActor, public IFlammable
{
    //other code
public:
    void Ignite() override;
}
```

On the calling side, if we had a flaming torch, we could add the following code to the collision handler for a collider around the flame particle:

Excerpt from possible flammable interface usage

```
if(UKismetSystemLibrary::DoesImplementInterface(otherActor,
    UFlammable::StaticClass()))
{
    IFlammable::Execute_Ignite(otherActor);
}
```

This code will let the torch ignite anything that is flammable and ignore everything that is not, without causing any errors or needing to do any unnecessary casting.

Interfaces, both in Blueprint and C++, offer us key benefits in keeping communication clean and efficient, allowing us to bundle up function names for use on implementing classes. Now that we've explored the benefits and the process to create both types of interface, we can look at another anonymous communication method – events.

Implementing event delegate communication across UE5

The other form of anonymous communication is the event delegate. Delegates are essentially function pointers that allow us to change the flow of logic dynamically at runtime. When invoked, they allow a signal to be sent to potentially several other parts of a program, without the sender knowing where the signal has gone. Using these as part of an event-driven approach is called an event delegate.

Event delegates can be compared to a radio station. The delegate exists as the station, transmitting a signal into the air when it is told to. Then, there are radios that can choose to subscribe to the station, receiving the signal. This forms a one-to-many relationship, with one-way communication. The radios cannot send messages back to the station, and the station does not know how many radios are tuned in.

Like when we explored interfaces, let's begin with the Blueprint implementation before moving on to the more complex C++ approach.

Event delegates in Blueprint

In Blueprint, delegates are made simple with one variant exposed. These are called Event Dispatchers. They do, however, function in the expected way.

Event Dispatchers need to be created in the class that will do the broadcasting (the radio station as such). We do this by clicking the + symbol in the **EVENT DISPATCHERS** dropdown of the **My Blueprint** tab.

Figure 6.11 – The Event Dispatchers rollout within the My Blueprint tab

An Event Dispatcher can have a series of inputs like any event in Blueprint. These are added at the top of the **Details** tab. You can also use the **Copy signature from** dropdown to select an existing event in the current Blueprint. This will duplicate all of the inputs from that event in the Event Dispatcher.

Figure 6.12 – Event Dispatcher variables and the Copy signature from dropdown

A Blueprint can respond to its own dispatcher in the same way that any listening delegate can respond. The dispatcher needs to be bound to an event, which then performs the desired response.

This is why, if you drag and drop the dispatcher from the Event Dispatchers list, onto the Blueprint graph, you will be presented with a popup containing **Call**, **Bind**, **Unbind**, **Unbind all**, **Event**, and **Assign**.

Figure 6.13 – The Event Dispatcher popup when dragged onto the Blueprint graph

These options are available wherever you choose to respond to a dispatcher (with the exception of an event). However, you will typically access them via the normal popup to create Blueprint nodes when dragging out from an object, which holds the dispatcher as a reference.

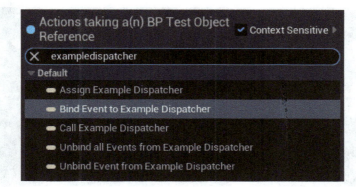

Figure 6.14 – The Blueprint context popup when searching for an Event Dispatcher by
name, after dragging from a reference to a class that holds the Event Dispatcher

Let's explore each of the options we are presented with from these popups so that you know when
and how to use them.

Call

The **Call** option activates the Event Dispatcher, sending the message to any subscribed delegates that
are listening for the event. This is typically used on the Blueprint that has the event, but can be called
from a separate Blueprint that has a direct reference to the actor with the dispatcher.

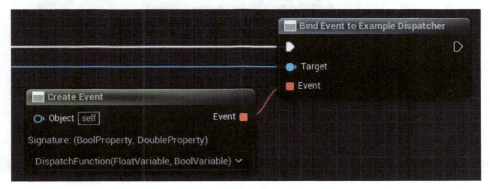

Figure 6.15 – A call dispatcher node with two variable inputs (Float and Bool)

The **Bind** option is used to identify which event should be run when the dispatcher is called; when
used, the associated event is added to the list of events associated with the dispatcher. In Blueprint, the
red line between the **Bind Event to** node and the **Event** node indicates which event it is to be bound to.

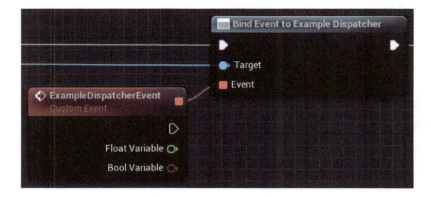

Figure 6.16 – A bind event node connected to a custom event that
will be run when the Event Dispatcher is called

This can be replaced by using a **Create Event** node, which can create a function or event with a matching signature (the same variables) or point toward an existing event. These are often used to try to keep the Blueprint graph tidy but are required to link to a function, as opposed to an event.

Figure 6.17–A bind event node connected to a Create Event node

Unbind and Unbind all

The **Unbind** node does the opposite of the bind node. When using **Unbind**, the event is removed from the list of events associated with the Event Dispatcher.

The **Unbind all** option does the same, but instead of just removing the linked event, it removes all events, nullifying the dispatcher from communicating with anything until another event is associated using a bind event node.

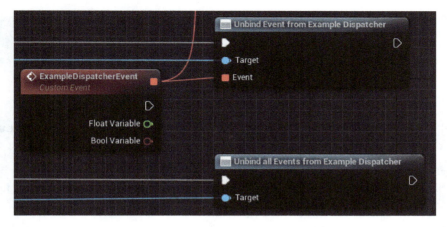

Figure 6.18 – The Unbind Event node, connected to a custom event, and an Unbind all Events node

The **Bind** and **Unbind** nodes allow us to dynamically link and unlink different events or functions to an Event Dispatcher, allowing us full control over what responds to a dispatcher at any point in the game.

Event and Assign

The **Event** and **Assign** options exist to help you create the nodes required to respond to an Event Dispatcher.

The **Event** option creates a **Custom Event** node with a matching signature, ready for you to connect to a bind event node later.

The **Assign** option creates both the Bind Event node and the matching custom event node, which can save a little bit of time when creating Blueprint logic.

Now that we have a grasp of the Blueprint implementation of events, let's move on to exploring the C++ implementation.

Event delegates in C++

In C++, event delegates are a little more complicated, but with that comes more control. For starters, there are a few different properties you can give each delegate before making events out of it. Let's take a look at them:

- DYNAMIC: This delegate can be serialized into the editor for easy Blueprint binding
- MULTICAST: This delegate can have multiple listeners at once and removes the ability to have a return value
- SPARSE: This delegate will not take up 1 byte in memory until it is subscribed to (this does make it slower to use, but it is more efficient for delegates that are rarely bound, such as a marketplace tool)

After deciding how the delegate should behave, you need to declare it as a new type. This is described by the following pattern in all caps:

```
DECLARE_<properties>_DELEGATE_<parameter amount>(<TypeName>, <arg1
type>, <arg1 name>... );
```

An example of this could be as follows:

```
DECLARE_DYNAMIC_MULTICAST_DELEGATE_TwoParams(FExampleSignature,
float, number, AActor*, actorRef);
```

This delegate will be serialized into the editor, with possibly multiple bound functions. Each bound function will have two arguments of type `float` and `AActor*`. This is the closest in behavior to a Blueprint Event Dispatcher.

There is also another type of delegate called an event. This is a version of a multicast delegate, where the first argument is the owning type, only that type can have events as members, and only the defining object can broadcast the events. With this extra security, you can expose events as public without fear that a different class will broadcast the event. However, with that comes extra design challenges to make your systems work in a clean manner.

> **Important note**
>
> From this point on, all event delegates used will be dynamic multicast to avoid confusion. This is not necessarily the best strategy, but it is a good starting point to then modify and restrict behavior later as needed.

That is how to define a new delegate type and create a new instance of a delegate from the type. To make delegates do something, you also need to know how to bind functions to them and fire them. This will change, depending on what properties were given to the event in it's declaration.

Working our way through the different properties that a delegate has, we can see that non-dynamic delegates have eight different functions that can bind listening functions to them for each single and multicast. It is recommended that you use the `BindUObject` function for single-cast delegates and the `AddUObject` function for multicast delegates. These both take a reference `UObject*` as a first argument and a function address as the second, in the style: `&ClassName::FunctionName`.

For dynamic delegates, there is less choice; single-cast dynamic delegates should use the `BindDynamic` function, and for multicast, it is advised you use the `AddUniqueDynamic` function, both with the same arguments as the non-dynamic delegates mentioned previously. The `AddDynamic()` function is ok to use with multicast dynamic delegates if you know the instance of the function has not already been bound to this delegate, otherwise it can be bound multiple times to the same event dispatcher.

Finally, to call delegates if they are single-cast, you should use the `ExecuteIfBound()` member function on the delegate, passing through arguments as necessary. For any multicast delegates, use the `Broadcast` member function. We will use a lot of multicast dynamic delegates when prototyping, as they offer the greatest flexibility to hook more listeners in as needed, leading us to make heavy use of the `Broadcast` function.

With the core concepts of delegates in C++ covered, let's move on to creating our own event delegate.

Building a useful delegate tool

Let's now look at building a useful class you can drop into any project. We will build on top of the damage system already present inside Unreal Engine. This utilizes a combination of an interface and the polymorphic nature of the engine, allowing you to deal damage to any actor and respond to damage from any actor. As an extension, we will look at making a Health Component that can sit on any actor, bind itself into this damage system automatically, and provide a couple of useful endpoints for any gameplay system to run with.

To start with, we need a new C++ class, so head to your C++ folder in the editor and make a new class based on **Actor Component**. Actor components are components for logic, as they have no associated rendering or transform within the world.

> **Useful tip**
>
> If you are using Rider 2022 or later, then a right-click on your module base folder will give you the option to **Add an Unreal Class**. In the popup, you can select an actor component, and it will function the same way but without the need for the engine to be running.

The first thing to think about is what delegate definitions we will need. The obvious one is for when the component has run out of health and needs to let the owning actor know that it has died. This could probably pass through a reference to the controller that instigated the chain so that, eventually, the game mode can attribute points to the correct controller:

```
DECLARE_DYNAMIC_MULTICAST_DELEGATE_
OneParam(FHealthDeadSignature, AController*, instigator);
```

The next useful piece of information to pass out to the component would be a signal whenever damage is dealt. There are a lot of options for what data to pass through. For instance, if the target is more of a *Call of Duty*-style game, you probably want to know what direction the damage was received from to feed that back to the player via some form of UI. In this instance, we will pass through information about the new and maximum health values, along with the change that just happened. This will let us update UI health bars and possibly spawn some damage numbers into the world:

```
DECLARE_DYNAMIC_MULTICAST_DELEGATE_
ThreeParams(FHealthDamagedSignature, float, newHealth, float,
maxHealth, float, healthChange);
```

We can then look at the header definitions. The only items that need to be public are the constructor and one of each delegate type declared previously. In the protected section, we need some float variables to track the current and maximum health. These are marked as protected so that the reflection specified in the UPROPERTY tag will work with the editor. Along with these, we'll use a `BeginPlay` override to set up some automatic behavior. Lastly, we'll add a handler function to the private section. This will be used to hook into the damage interface system already present inside the actor this component is attached to. With that all written out, the class definition should look a little like this:

HealthComponent.h class definition

```cpp
UCLASS(ClassGroup=(Custom), meta=(BlueprintSpawnableComponent))
class <module name>_API UHealthComponent : public UActorComponent
{
    GENERATED_BODY()

public:
    UHealthComponent();

    UPROPERTY(BlueprintAssignable)
    FHealthDeadSignature OnHealthDead;

    UPROPERTY(BlueprintAssignable)
    FHealthDamagedSignature OnHealthDamaged;

protected:
    UPROPERTY(EditAnywhere, BlueprintReadWrite)
    float _MaxHealth;

    UPROPERTY(VisibleAnywhere, BlueprintReadOnly)
    float _CurrentHealth;

    virtual void BeginPlay() override;

private:
    UFUNCTION()
    void DamageTaken(AActor* damagedActor, float damage,
        const UDamageType* dmgType, AController* instigator,
        AActor* causer);
};
```

After generating the definitions for these functions, we can add a default value for maxHealth into the constructor of maybe `100.f`, but this line isn't too important. The `BeginPlay` override needs to bind our private DamageTaken function to the OnTakeAnyDamage event of the owning actor it is attached to, as well as setting the initial value for _CurrentHealth:

HealthComponent.cpp Constructor and BeginPlay

```cpp
UHealthComponent::UHealthComponent()
{
    _MaxHealth = 100.f;
}


void UHealthComponent::BeginPlay()
{
    Super::BeginPlay();
    _CurrentHealth = _MaxHealth;
    GetOwner()->OnTakeAnyDamage.AddDynamic
        (this, &UHealthComponent::DamageTaken);
}
```

Then, the logic for how damage is taken and how the events are fired will all go into the DamageTaken function. Note the use of FMath::Min to make sure the damage received never goes beyond the amount of health remaining. This could be easily altered to allow for *Doom*-style forgiveness mechanics, where the player cannot be taken from >1% health to 0% in one hit and can only die from 1% health. Alternatively, it could be removed to allow for overkill calculations if that matters for gameplay:

HealthComponent.cpp DamageTaken function

```cpp
UHealthComponent::DamageTaken(AActor* damagedActor,
    float damage, const UDamageType* dmgType,
    AController* instigator, AActor* causer)
{
    float change = FMath::Min(damage, _CurrentHealth);
    _CurrentHealth -= change;
    OnHealthDamaged.Broadcast(_CurrentHealth, _MaxHealth,
    change);
    if(_CurrentHealth == 0.f)
    {
        OnHealthDead.Broadcast(instigator);
    }
}
}
```

To make use of this component, you can now go to any actor and add this component to its hierarchy. It will automatically bind itself, and we have built in two events that can be bound to Blueprint and C++ functions.

In the example project, add the health component to the enemy unit **Blueprint Editor**. Create a listener for the **OnHealthDead** event by selecting the new component in the **Hierarchy** panel and clicking the plus button next to the event in the **Details** > **Events** panel.

Link up the Blueprint functions in the red comment block marked **Handle Unit Death** as shown in *Figure 6.19*, and you will see that because the player has been built with the damage system in mind, you can now move about and destroy enemy units.

Figure 6.19 – The fully linked up event

This functional example shows how event delegates can be used to communicate anonymously, thereby removing the need for object references everywhere. Using this communication method effectively helps to reduce the coupling of code bases, making them easier to maintain over time as more are added.

Summary

This chapter covers the two main methods of cleaning up communication in a code base. We have covered functional examples of how to use both interfaces and event delegates within our C++ setup, as well as the technical setup, using the U and I prefixes of an interface properly, and what each of the characteristics of a delegate mean. Now that the functional understanding is sorted, the next step should be to practice with these tools to get a better understanding of how they affect the code base. We will do this in the next chapter by looking at how we deploy both interfaces and event delegates with function calls, achieving a perfectly decoupled system where communication is as anonymous as possible. This naturally creates modular code design, allowing for much easier cohesion within teams.

7

A Perfectly Decoupled System

In the previous chapter, we learned about interfaces and event delegates and how they can be used as tools to remove the need for unnecessary object references within our program structures. This was the concept of decoupling.

In this chapter, we will take the concept of decoupling and extrapolate it throughout our system design. We will look at a structure that should provide a robust framework for any system and the standard method for planning it.

In this chapter, we'll cover the following main topics:

- Using UML to plan a sample hierarchy
- Decoupling the reference train

By the end of this chapter, you will be able to architect complex game communication hierarchies in a fully decoupled and modular way. This will allow you to build more maintainable and expandable systems that enable development team cohesion.

Technical requirements

This chapter will be focused on implementing the tools from *Chapter 6* in the same project, linked on GitHub.

If you want to jump straight in at this point, you can download the *chapter7* branch from GitHub at https://github.com/PacktPublishing/Game-Development-Patterns-with-Unreal-Engine-5/tree/main/Chapter07

Otherwise, the planning section only really requires a pen and paper or a free drawing tool. Once the purpose and vocabulary of UML diagrams have been grasped, there are helpful online tools for speeding up their generation. The easiest to use is Mermaid (https://mermaid.live), which provides a text editing interface for diagram editing. It presents as a paid service if you wish to save multiple diagrams, but seeing as every element of your diagram or chart is serialized as text in the URL, shown in the address bar of your browser, a simple copy and paste of the URL into a notes app will save the diagram. Other tools include Photoshop and Paint or dedicated drawing tools such as drawio. com (`https://www.drawio.com/`) and whimsical.com (`https://whimsical.com/`).

Using UML to plan a sample hierarchy

Unified Modeling Language (**UML**) is a tool that programmers use to visually communicate the structure of large systems during the design phase and when creating technical documentation. Classes are represented as boxes, with relationships shown through different types of arrows linking the boxes.

What are the types of class relations?

Let us take a look at different types of class relations and how they are presented using UML class diagrams.

Inheritance

Standard arrows, as seen in *Figure 7.1*, show inheritance with the parent being indicated by the end with the arrow head.

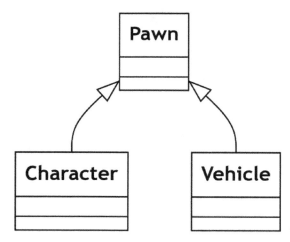

Figure 7.1 – An example of a UML class diagram showing inheritance syntax

In *Figure 7.1*, the **Character** and **Vehicle** classes inherit from the **Pawn** class.

Composition

Composition is a type of class relation where the composed classes give the parent class the functionality. This tool is seen a lot with multi-functional classes being broken down into their separate functions then composed together into a wrapper. You can show composition with a solid diamond at the end of a line.

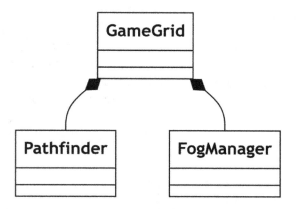

Figure 7.2 – An example UML class diagram showing composition syntax

In *Figure 7.2*, **GameGrid** composes a reference to **Pathfinder** and **FogManager**.

Aggregation

Aggregation is a concept crucial to designing decoupled systems as it quite literally represents the coupling that is present. At some point in the communication chain, for objects to communicate, at least a one-way connection must be established. Aggregation is when that connection is between two non-dependent classes. Similar to composition, aggregation is also shown by a line with a diamond at the head but the diamond for aggregation is just an outline.

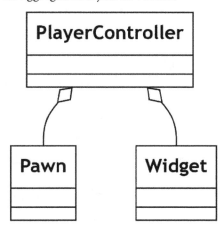

Figure 7.3 – An example UML class diagram showing aggregation syntax

Figure 7.3 shows **Pawn** and **Widget** aggregated into **PlayerController**. Both can exist and function without this connection so it is an aggregation.

What is a sequence diagram?

In addition to the UML class diagrams we've explored so far, UML has more forms it can take. A **sequence diagram**, shown in *Figure 7.4*, shows the execution of functions between objects over time. This is a crucial way to show how signals get around your class structures. Using a sequence diagram after something goes wrong can help identify what a call stack should look like at different lines of execution to pinpoint errors.

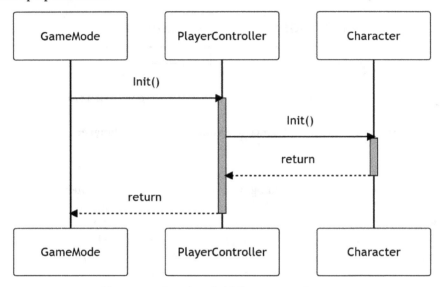

Figure 7.4 – An example UML sequence diagram

In *Figure 7.4*, we are communicating an initialization chain from **GameMode** through **PlayerController** to **Character**. The sequence diagram helps us see that when the **Character Init** function is executing, our call stack should be three deep and what order each class is in the chain. These diagrams are necessary for clarifying key communications but, as with all UML, it can have the opposite effect if it tries to encompass too much at once.

With these tools in our belt, we can better describe the structures coming up in this chapter to other developers. The benefit of using UML over coding segments is in its standardization and agnosticism, meaning the programmer we show this to doesn't need to speak the same language to understand what we mean.

Decoupling references is a key skill for all developers in all object-oriented languages to understand. With universal methods of communication such as UML, we can establish a precedent and enforce it across all levels of development. Let's now use loads of sequence diagrams and aggregation to fix some messy, heavily coupled code.

Decoupling the reference train

Now let's take knowledge of these three connection types and see how we can associate every class in the example UML class diagram that follows, using a few rules. Once everything is associated, we'll implement decoupling in an example and look at the benefits this brings.

Modularity and decoupling

We'll start with a bunch of scattered classes, which are organized in a way that will function but is messy to work with, as shown in *Figure 7.5*. To make everything function here clearly, the developer has added references, as and when needed, to any class.

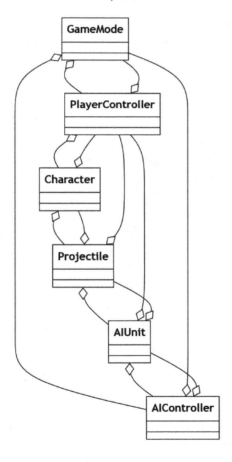

Figure 7.5 – Example UML showing a messy system

Let's analyze how the **Character** to **Projectile** interaction occurs by looking at the sequence diagram in *Figure 7.6*. When the character *fires*, it spawns a projectile but then doesn't hold onto the reference. Instead, the projectile grabs a reference to the character and when it hits an AI unit, this reference is used to feed back whether it killed anything through a function call. Why is this bad? It's not necessarily going to impact the final built game, but it does make the code involved in this interaction single-use. If an enemy unit wants to fire, due to it being a different setup, we are left with two bad options:

- Bastardization through inheritance and overloading functions, thereby breaking several of our SOLID principles

- Repeat ourselves in a new actor, which, as we know, should ring alarm bells

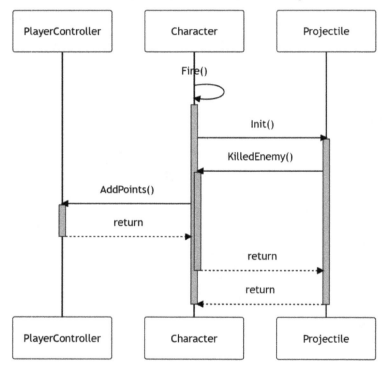

Figure 7.6 – Example UML sequence diagram showing the complexity of a messy interaction

The solution to this problem is modularity and decoupling. We need to establish a hierarchy where each class is connected to a single tree, where connections represent ownership. The first step is to analyze class responsibilities. The **Character** class currently takes care of too many areas. It deals with player representation in the world but also weapon logic. Due to this logic possibly being duplicated in the AI enemy class, it makes sense to spin it out into its own class that can be shared. Linking this into an aggregation chain with the **GameMode** at the top, flowing down through the **Character** and **Weapon** to the **Projectile**, gives us the tree we were aiming for from the start, as shown in *Figure 7.7*:

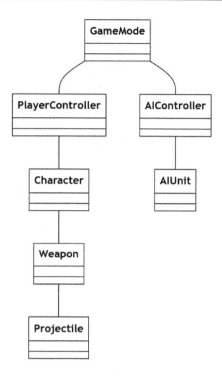

Figure 7.7 – A UML class diagram showing classes in a tree structure

In general terms, the tree we established in *Figure 7.7* is conceptual. There could be pointers to owned classes in parents for practicality, but it mostly exists in the design document. In Unreal, everything below the PlayerController can be spawned at runtime, which helps to make the structure more concrete. All spawned objects can be cached as `TObjectPtr<T>` variables, with `T` being the exact type we need. To link the tree up above the **Character** level, it gets more specific. When a new level loads, it creates new Controllers automatically, but it does so using multiplayer terminology and a networked method. Overriding the `PostLogin` and `Logout` functions of `AGameModeBase`, PlayerControllers can be cached in a `TArray<TObjectPtr<APlayerController>>` for later use. If you want to make a distinction between AIController and PlayerController in the structure, then you will need to store them separately. PlayerControllers can be cached via the `PostLogin` function with a cast to the respective base type as needed, but AIControllers must be cached when the AI-controlled pawn is spawned and accessed via the instigator reference.

Important note

Be aware that `PostLogin` only fires for PlayerControllers, yet `Logout` deals with all controller types, including AI. This makes defensive coding important in `Logout`.

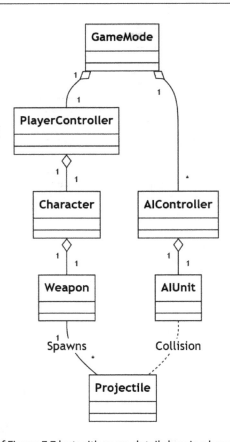

Figure 7.8 – Repeat of Figure 7.7 but with more detail showing how classes are connected

Figure 7.8 shows how the different methods described in this section could be used to connect the classes in memory.

Establishing infrastructure

The next step is to start establishing infrastructure, such as public functions, event delegates, and interfaces, to provide entry and exit points for signals. Each fulfills a very specific purpose:

- **Public functions** are used for communication when there is a cached variable reference of the correct type. Following the aggregation lines that represent this connection, as shown in *Figure 7.8*, all function calls will happen to head down the tree through the chain of ownership.

- **Event delegates** allow anonymous communication when the listener has a cached variable reference for the event's class. This ends up being the exact opposite of the public function calls, allowing communication to back up the reference chain toward the top of the tree.

- **Interfaces** are the final tool and allow communication between different branches without creating a link. Providing an `AActor*` reference that can be gained at runtime, usually via some kind of collision event, the interface can be used to invoke some kind of function without needing to know the exact type of the receiver.

This boils down to a simple communication rule: *functions down, events up, and interfaces sideways.* The result is a decoupled system in which each class only has reference to the one layer below it and yet signals can be sent all over with ease.

Implementing decoupled design in an example

The best way to embed decoupled design into your process is to practice it. As a start, we will focus on how using the player character in our example project can receive points shown on the UI for eliminating enemy units. We will start with the UML class diagram shown in *Figure 7.9*:

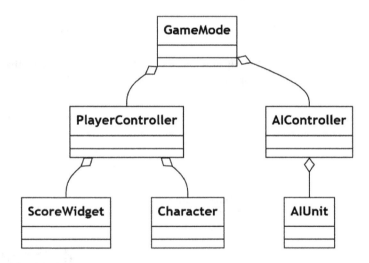

Figure 7.9 – Planned UML class diagram for the linked project

Next, we will overlay the signal path onto the UML, as shown in *Figure 7.10*:

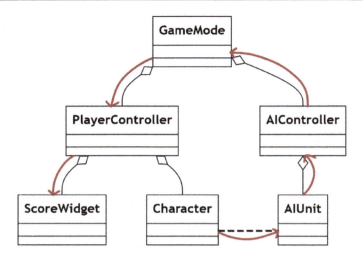

Figure 7.10 – Signal path overlayed on UML class diagram

Overlaying the signal path onto the UML will allow us to plan the infrastructure effectively, producing a UML sequence diagram, as shown in *Figure 7.11*:

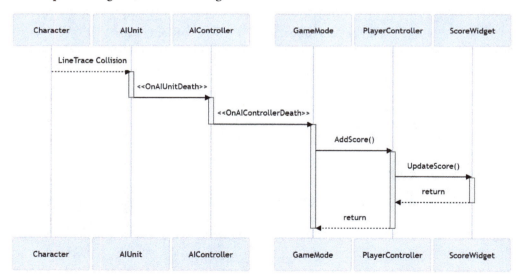

Figure 7.11 – UML sequence diagram showing the signal path for killing an enemy

To put this decoupled plan into practice, head to the folder labeled *Chapter 7* in the example project. Open the level in this folder and take the following steps. Whenever a class is mentioned, please use the *Chapter 7* version denoted by _CH7 in the class and filenames. All of the following code is for you to add and embellish as you see fit. All the needed classes have been created already. This exercise only requires you to add the new code to the areas specified by the protection-level tags. Any new variables will need to be linked up in the inspectors of the Blueprint children in the editor, but this is made clear at the end of this section.

Stepping through the implementation, let's start at the top in the game mode. GameMode_CH7 will have links to PlayerController_CH7 and any AI spawned in the level. This may work better with an AI manager for larger games, but we have no need to over-complicate the process. These references are cast to the correct types and cached. After casting, listener functions can be attached to the event delegates on the controllers, establishing the first link, but we haven't made any controllers yet so we will return here once the initialization chain is complete.

The GameMode header is a simple set of holder variables and the start of the initialization chain in a BeginPlay override.

Let's get started with the header file.

GameMode_CH7.h excerpt

```
class AGameMode_CH7 : public AGameModeBase
{
public:
    virtual void PostLogin(APlayerController* NewPlayer) override;
    virtual void Logout(AController* Exiting) override;

protected:
    UPROPERTY(VisibleAnywhere, BlueprintReadOnly)
    TObjectPtr<APlayerController_CH7> _PlayerController;

    UPROPERTY(VisibleAnywhere, BlueprintReadOnly)
    TArray<TObjectPtr<AAIController_CH7>> _AIControllers;

    virtual void BeginPlay() override;
};
```

The GameMode body deals with PlayerController initialization and reference registry in the PostLogin and Logout functions as that is what Unreal will automatically call when PlayerControllers enter and exit the game. This can be confusing as it is networking terminology but Unreal still uses it for single-player games. On the other hand, AIControllers are gathered and initialized on BeginPlay as they are assumed to be in the level already.

So, now, we can set these up in the body file.

GameMode_CH7.cpp excerpt

```cpp
void AGameMode_CH7::PostLogin(APlayerController* NewPlayer)
{
    if (APlayerController_CH7* castedPC = Cast<APlayerController_CH7>
        (NewPlayer))
    {
        _PlayerController = castedPC;
        _PlayerController->Init();
    }
    Super::PostLogin(NewPlayer);
}

void AGameMode_CH7::Logout(AController* Exiting)
{
    if (Exiting == _PlayerController)
    {
        _PlayerController = nullptr;
    }
    Super::Logout(Exiting);
}

void AGameMode_CH7::BeginPlay()
{
    Super::BeginPlay();
    TArray<AActor*> outActors;
    UGameplayStatics::GetAllActorsOfClass
        (GetWorld(), AAIController::StaticClass(), outActors);

    for (AActor* actor : outActors)
    {
        _AIControllers.Add(Cast<AAIController_CH7>(actor));
    }

    for (AAIController_CH7* ai : _AIControllers)
    {
        ai->Init();
    }
}
```

Next, the controllers need to gain reference to and initialize the characters they are controlling. This is a simpler process than before as the controllers all spawn in their characters via an initialization chain. This means there is no need for casting; the characters are already the correct type. This also applies to the `ScoreWidget` class we will use to display the player's score, which has been provided as part of the `Chapter Resources` folder.

The PlayerController header has a custom public function for initialization, a class reference to the type of pawn it would like to spawn, and an object reference to the actual pawn it has spawned.

Write out the PlayerController header file as shown in the following code:

PlayerController_CH7.h excerpt

```cpp
class APlayerController_CH7 : public APlayerController
{
public:
    void Init();

protected:
    UPROPERTY(EditAnywhere)
    TSubclassOf<APawn> _PlayerPawn;

    UPROPERTY(VisibleAnywhere, BlueprintReadOnly)
    TObjectPtr<ACharacter_CH7> _Character;
};
```

In the PlayerController body, we use the functions provided to us by Unreal within the GameMode class to spawn a new pawn of the type specified in the header at one of the player start points. The PlayerController then possesses it and calls the initialization function on that pawn.

Let's set that up.

PlayerController_CH7.cpp excerpt

```cpp
void APlayerController_CH7::Init()
{
    if (GetPawn() != nullptr)
    {
        GetPawn()->Destroy();
    }

    UWorld* const world = GetWorld();
    AActor* tempStart = UGameplayStatics::GetGameMode
        (world)->FindPlayerStart(this);

    FVector spawnLoc = tempStart != nullptr ?
        tempStart->GetActorLocation() : FVector::ZeroVector;
```

```
    FRotator spawnRot = tempStart != nullptr ?
        tempStart->GetActorRotation() : FRotator::ZeroRotator;

    FActorSpawnParameters spawnParams;
    spawnParams.SpawnCollisionHandlingOverride =
        ESpawnCollisionHandlingMethod::AdjustIfPossibleButAlwaysSpawn;

    APawn* tempPawn = world->SpawnActor<APawn>
        (_PlayerPawn, spawnLoc, spawnRot, spawnParams);
    Possess(tempPawn);

    if (ACharacter_CH7* _Character = Cast<ACharacter_CH7>(tempPawn))
    {
        _Character->Init();
    }
}
```

Down the other branch from the GameMode, our AIController header is similar to the PlayerController minus the reference to the class it would like to spawn, as with a standard Unreal AI setup the pawn is already in the world.

So, the header file for the AIController needs to be set out like this:

AIController_CH7.h excerpt

```
class AAIController_CH7 : public AController
{
public:
    void Init();

protected:
    UPROPERTY(VisibleAnywhere, BlueprintReadOnly)
    TObjectPtr<AAIUnit_CH7> _Unit;
};
```

This point of the AIController being a generally simpler implementation extends to the initialization function. As there is no need to spawn any new pawns, we can simply get a reference to the currently possessed one and call its initialization function, passing the communication on.

Write the Initialization function as shown:

AIController_CH7.cpp excerpt

```
void AAIController_CH7::Init()
{
    if (APawn* pawn = GetPawn())
```

```
    {
        if (_Unit = Cast<AAIUnit_CH7>(pawn))
        {
            _Unit->Init();
        }
    }
}
```

The last initialization is the AIUnit class. In the header, we have a variable to hold a HealthComponent, like the one we made at the end of *Chapter 6*, and a listener function to bind to the component's event for death.

So, let's setup the AIUnit header.

AIUnit _CH7.h excerpt

```cpp
class AAIUnit_CH7 : public APawn
{
public:
    AAIUnit_CH7();
    void Init();

protected:
    UPROPERTY(VisibleAnywhere, BlueprintReadOnly)
    TObjectPtr<UHealthComponent> _HealthComp;

    UFUNCTION()
    void Handle_HealthDeath(AController* causer);
};
```

The body makes a new instance of a HealthComponent in the constructor and binds the listener function in the initialization.

We should now add the HealthComponent and initialization functions to the AIUnit body file.

AIUnit _CH7.cpp excerpt

```cpp
AAIUnit_CH7::AAIUnit_CH7()
{
    _HealthComp = CreateDefaultSubobject<UHealthComponent>
    (TEXT("Health"));
}

void AAIUnit_CH7::Init()
```

```
{
    _HealthComp->OnDead.AddDynamic
        (this, &AAIUnit_CH7::Handle_HealthDeath);
}
```

Now that all classes are initialized, we can step back up through the chain, adding the event delegates and their `listener` functions, starting with how the `AIUnit` responds to the health component death event. This needs to be done on both branches of the tree for the player and AI side, but we will only show the AI side as that is where the points come from. The listener should broadcast the `AIUnit` death event for now, including however many points this `AIUnit` was worth. This link is where animations and sound effects could be played to provide feedback to the player.

> **Extension task**
>
> The other damage event in the `HealthComponent` is being ignored for the purpose of explaining the process. As an extension task after finishing the chapter, try to hook this up so that when an AI character receives damage, it feeds through the chain to the controller and updates the health bar from there.

Adding the following code through the `AIUnit` header, we declare a new delegate type called `FAIUnitDeathSignature` with two parameters. This type is then used in the class header to make the `OnUnitDeath` public delegate. We also add an integer here so that we can balance the value of destroying the unit to the player.

So, lets add those to the header file.

AIUnit _CH7.h excerpt

```
DECLARE_DYNAMIC_MULTICAST_DELEGATE_TwoParams
    (FAIUnitDeathSignature, AController*, causer, int, points);

class AAIUnit_CH7 : public APawn
{
public:
    UPROPERTY(BlueprintAssignable)
    FAIUnitDeathSignature OnUnitDeath;

protected:
    UPROPERTY(EditAnywhere, BlueprintReadWrite)
    int _PointValue;
};
```

The `AIUnit` body is much simpler though, as we are just using it to be a link in a chain, so once the `HealthComponent` death listener function fires, it can broadcast the new `OnUnitDeath` delegate and destroy the pawn.

Add the following code to the body file.

AIUnit _CH7.cpp excerpt

```
void AAIUnit_CH7::Handle_HealthDeath(AController* causer)
{
    OnUnitDeath.Broadcast(causer, _PointValue);
    Destroy();
}
```

Following up the chain, the AIController can start listening to the AIUnit's pass-through death event. This level is another pass-through for now, but could be where the controller recycles itself and finds another AIUnit to spawn depending on the gameplay loop.

The AIController header gets a similar set of items to the AIUnit with a new delegate type definition that takes two parameters. We don't reuse the definition in the unit so that when we come to expand the game, these two classes aren't tied to each other and can be swapped out as necessary. There is also a private function that matches the signature for the AIUnit OnUnitDeath delegate so it can be added as a listener.

Go ahead and add the following code to the header file.

AIController_CH7.h excerpt

```
DECLARE_DYNAMIC_MULTICAST_DELEGATE_TwoParams
    (FControllerDeathSignature, AController*, causer, int, points);

class AAIController_CH7 : public AController
{
public:
    UPROPERTY(BlueprintAssignable)
    FControllerDeathSignature OnControllerDeath;

protected:
    UFUNCTION()
    void Handle_UnitDeath(AController* causer, int points);
};
```

The AIController body binds the listener function to the AIUnit OnUnitDeath delegate right after the Init function is called. This way, we know that all variables inside the AIUnit have been set up correctly before we start listening for gameplay signals. The listener function itself is then just a passthrough broadcasting the OnControllerDeath delegate.

Let's add that to the code in the body file.

AIController_CH7.cpp excerpt

```
void AAIController_CH7::Init()
{
    if(APawn* pawn = GetPawn())
{
if(_Unit = Cast<AAIUnit_CH7>(pawn))
        {
            _Unit->Init();
            _Unit->OnUnitDeath.AddDynamic(this,
                &AAIController::Handle_UnitDeath);
        }
    }
}


void AAIController_CH7::Handle_UnitDeath(
    AController* casuer, int points)
{
    OnControllerDeath.Broadcast(causer, points);
}
```

The GameMode can now link a listener function to all the AIControllers. This function just calls a `public` function on the PlayerController to increment its points for now, but it can be used as a jumping-off point to check against win conditions, and in a multiplayer scenario, make sure all players have an updated scoreboard via the `GameState`.

This following excerpt shows where to add the listener function declaration in the GameMode header.

GameMode_CH7.h excerpt

```
class AgameMode_CH7 : public AgameModeBase
{
protected:
    void Handle_ControllerDeath(AController* causer, int points);
}
```

In the GameMode body, we then bind the listener function to each AIController right after we call the `Init` function on each one. This time, the listener function doesn't simply broadcast another event, as this is where the communication reverses and starts going down another branch of the communication hierarchy that we designed earlier in the chapter. Instead, we call a function on the PlayerController telling it to add points. This is not strictly the best way of doing this, but it does show the structure cleanly. Ideally, you would check which controller needed the points based on the `causer` argument passed through the delegate chain.

So, add the following code to the body file:

GameMode_CH7.cpp excerpt

```
void AGameMode_CH7::BeginPlay()
{
    … other code from before …
for(AAIController_CH7* ai : _AIControllers)
{
    ai->Init();
    ai->OnControllerDeath.AddDynamic(this,
        &AgameMode_CH7::Handle_ControllerDeath);
}
}

void AGameMode_CH7::Handle_ControllerDeath(
    AController* causer, int points)
{
    _PlayerController->AddPoints(points);
}
```

The Player controller then needs to implement this `public` function so that when the game mode tells it to add points, it updates the UI as well. The UI references are included in the class for you already – that's why they are missing from the header in the following code.

The PlayerController header just needs the public accessor `AddPoints` function and a variable to store the current number of points. This is arbitrary but they must be stored somewhere, and this seems like as good a place as any to store them.

So, let's add these:

PlayerController_CH7.h excerpt

```
class APlayerController_CH7 : public APlayerController
{
public:
    void AddPoints(int points);
protected:
    int _Points;
}
```

`AddPoints` does what it says on the tin, but also serves as a signal passthrough for the UI where we tell it to update the value shown onscreen with the new points value.

Add the following excerpt to the body file:

PlayerController_CH7.cpp excerpt

```cpp
void APlayerController_CH7::AddPoints(int points)
{
    _Points += points;
    _PointsWidget.UpdatePoints(_Points);
}
```

With that all done, you should be able to play the game as before from this new level, but when you destroy enemy AI units, the number at the top left of the viewport increases each time. This is the core of a game loop built in a decoupled way. Of course, our AI needs to fight back and there needs to be some balancing to make this into a good gameplay experience, but functionally it is all there.

Benefits of decoupling

The benefit of setting things up in a decoupled way should now be evident. If we need to swap out game modes for a different objective-based mode, then we can do that as no class needs a direct reference to the game mode. As long as all the same functions are called and the events are listened to, everything will function. You can also check the size of the dependencies for each class using the method shown in *Chapter 1* to see how small each class is in memory now that it has a reference to only one layer below itself and maybe an interface or two.

Summary

With this chapter completed, you should be equipped to design communication hierarchies for games with a focus on anonymous modular design in UML for implementation within Unreal. We have covered the basics of UML and why it is useful as a planning and communication tool. Using this UML, we then set about taking a simple communication and anonymizing it using the event delegate tool from *Chapter 6* to decouple the reference chain as much as we could. This anonymous modular should work for most communications you design from here on, with exceptions being extremely rare.

In the next chapter, we are going to look at patterns you can set up as a library to move around with you between projects. We will look at why you should not overuse the singleton pattern, as most people do, and why you should make use of the command and state patterns in almost every project.

Part 3:
Building on
Top of Unreal

In this part, we will take the leap from looking at patterns that have already been scaffolded to making some of our own.

Each chapter deals with a different category of pattern, starting with behavioral patterns that allow classes to serve their use more cleanly, through structural patterns that assist you when building systems for a large team to work on, to optimization patterns that aim to speed up your code. By the end of this section, you should have a host of patterns built in a modular way, ready to be transferred between projects.

This part has the following chapters:

- *Chapter 8, Building Design Patterns – Singleton, Command, and State*
- *Chapter 9, Structuring Code with Behavioral Patterns – Template, Subclass Sandbox, and Type Object*
- *Chapter 10, Optimisation through Patterns*

8

Building Design Patterns – Singleton, Command, and State

Whereas the previous chapter looked at a methodology of code architecture design, this chapter will look at three design patterns you can build yourself that have applications across many game genres.

The patterns being covered are as follows:

- Singleton pattern – understanding why it's a Pandora's box that often gets overused
- Command pattern – how it has many uses beyond the obvious
- The many levels of state machine traveling down the rabbit hole and seeing how far we can push its concept

The aim of this will be to make some base classes that can be imported into any future project to speed up development. By the end of the chapter, you should understand why so many online resources overuse the Singleton pattern, what a hidden gem the Command pattern is, and how deep customization can go with the humble state machine.

Technical requirements

This chapter will be building on the **real-time strategy** (**RTS**) framework project of previous chapters using the `chapter8` branch on GitHub, which can be downloaded from `https://github.com/PacktPublishing/Game-Development-Patterns-with-Unreal-Engine-5/tree/main/Chapter08`

Implementing a Singleton pattern – understanding why it's a Pandora's box

The official point of the Singleton pattern is to ensure there is only one instance of a class in existence at any one time, hence the name "single"-ton. Unfortunately, this often gets packaged and confused with a public static variable to this one existing object. The actual idea of only having one object instance of a class makes sense. You might have a manager that needs to exist in every level, but if you don't know the path the player took to get to this level, then you have no idea if one has been spawned yet. The solution is to make the manager a Singleton class and have a copy at every level. We can do this with the following code, using a static variable pointing to the one that exists:

Example Singleton.h excerpt

```
UCLASS()
public Singleton : public AActor
{
    static TObjectPtr<Singleton> _instance;

public:
    void Init();
}
```

We then have a choice on whether we delete the new one if it is a second or if it should assume the position of the instance, deleting what was previously there . The body file therefore would be:

Example Singleton.cpp excerpt

```
void Singleton::Init()
{
    if (_instance == nullptr)
    {
        _instance = this;
    }
    else
    {
        this->Destroy();
    }
}
```

There is an innate question arising from this. Internally, our monologue could sound like: "If only one instance of this class will ever exist, then why does it need to be instanced in the first place? Surely, making the entire class static has the same effect and will take up less memory with the functions and variables only existing on the stack, negating the need for a Singleton pattern." You could answer that with a situation. Perhaps there is a need to replace the object in the instance slot with every new level. That is one argument for a Singleton pattern, but the usage is limited and can usually be designed around using the pattern we discussed at great length in *Chapter 7*.

The other issue is that most implementations of the Singleton pattern use a public static variable for tracking the instance. This has led many people to think the purpose of this pattern is for an easy communication link between a top-level system and any object that needs it. Doing this potentially couples every class with the Singleton pattern, which we have established in previous chapters is a thing to be avoided. The effect can be seen clearly in *Figure 8.1*, where every class aggregates the **Singleton** class into its memory footprint because it has one or more references within:

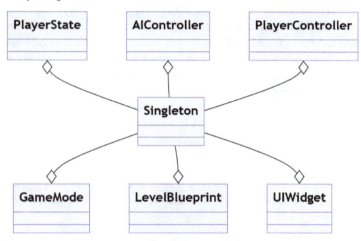

Figure 8.1 – UML diagram of a Singleton pattern being aggregated into lots of classes

The correct usage is a private static variable used to track the instance, as shown at the start of this section. Then, with the instance created by a manager, it fits into the hierarchical class tree shown in *Chapter 7*. This manager can then distribute the direct reference to any classes under it via the initialization chain. This is what we know as **dependency injection**. This does still create coupling, but the spread is much more controlled. An example custom initializer function for a unit in a grid-based game could have the dependency of the grid injected into it instead of making the grid a Singleton pattern that anything can access. Next, you can see an example of how this could be constructed and how little this method impacts the flow of the code:

Dependency injection example

```
void AUnit::Init(AGrid* grid)
{
    _GridRef = grid;
}
```

```
void AUnit::Move(FVector2 targetCoords)
{
    Path p = GridRef->GetPath(_CurrentCoords,
        targetCoords))
    if(p != nullptr)
{
        //Move the Unit
    }
}
```

Overall, the actual concept and code for the Singleton pattern is relatively simple; the problem lies in how it is used. Even with the cleaner implementation, most – if not all – cases where a Singleton pattern has been used can be replaced with a different pattern to make the code easier to expand. Aggregation may mean that this functionality can become a component, or there may need to be a shift in the structure to something more like a subclass sandbox pattern (to be discussed later, in *Chapter 9*) where the functionality is statically defined in a parent. Dependency injection takes most of the replacement duty, as the common use for a Singleton pattern will be a global utility class such as a **Fog of War** manager on a tiled grid. This is better sent through the initialization chain as a dependency injection so that other classes don't also have access allowing other developers to incorrectly use the functionality.

Now that the temptation of the Singleton pattern has been expunged from our minds, we can move on to patterns that are useful in multiple scenarios, such as the Command pattern.

Implementing the Command pattern for different use cases

The Command pattern adds a layer of separation between the request for an action and that action being carried out. The implementation looks like what is shown in *Figure 8.2*, where the Command class parent is abstract and only has a constructor, execute(), and undo() functions that all take no arguments. The idea is that the child classes are more specific and contain all the object references needed to execute properly:

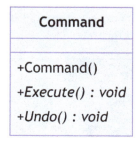

Figure 8.2 – UML diagram showing the structure of a Command pattern base class

The purpose of a command is to reify the abstract idea of an action so that we can store it in a list. This list can have many uses, but the most identified is the undo queue that Microsoft made synonymous with its keyboard shortcut, *Ctrl + Z*. When an action is performed, a Command object of the relevant type is created and added to the list. The command is executed and left in this list until it falls off the back; this keeps the list in chronological order. If the user presses the undo key, then the last command in the list has its undo function called, and a pointer for the most recent command moves back one. This is shown in *Figure 8.3*, where you can see that creating a new command after some have been undone chops the undone commands off and inserts the new command as the head of the list:

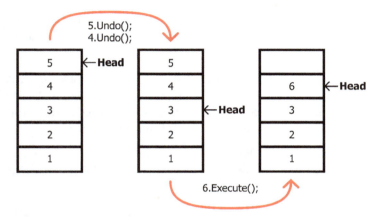

Figure 8.3 – Diagram showing the process of adding, undoing, and branching commands

In games, we can use this functionality for a plethora of different scenarios. The most common are strategy games, where the player can queue up actions for their units to be executed in a certain order, each only executing after the last has finished. Surprisingly few strategy games make use of the Command pattern's undo capabilities, but that may be because it would reduce the cost of actions. Undo is making more of an appearance in modern games in the form of a rewind mechanic; exploited in classics such as *Braid* and incorporated as a feature into AAA games such as the *Forza Horizon* series. This real-time application has varying methods of execution; it is likely that *Braid* simply records all moving objects' positions on a timer as there is so little going on in the scene. This approach clearly wouldn't work for a racing game with complex physics; instead, you could record each time the player's input changes as a different command. The undo queue then has to run the game backward and understand when commands were added to the list to preserve momentum.

Saving each of the players' input actions into a list can also work as a sort of replay-saving mechanic. This works for games such as the *Trials* series and fighting games such as *Super Smash Bros.* because there are no elements of randomness in the mechanics. If you input the same action at the same time, you will have the same result every time. This could lead to large replay files, with each input axis saving potentially a value every frame. There are potential ways around this, such as only saving the result of these inputs when an action is performed that relies on the state of the driven element. This could mean instead of saving every mouse movement in *World of Tanks*, the barrel rotation and time are saved every time a shell is fired and the barrel is hit. This way, all the little motions that add to zero are ignored from the replay file.

Command pattern for undo functionality in Blueprint Utilities

In Unreal, we can make tools to use in the editor called **Blueprint Utilities** (formerly known as Blutilities during early development). Blueprint Utilities can be created as either right-click menu actions or as Utility Widgets that operate in a floating window or can be docked into the UI and can manipulate assets (the files in the content browser) or actors (elements within the world) to complete repeated actions, to reduce the impact of workflow steps or simply to remove the requirement for user input to reduce opportunities for mistakes.

Editor Utilities can hold any number of tools within them, with each tool being created as its own function, having its own graph or a custom event within the standard Blueprint Event Graph. Actor and Asset Utility Widgets require us to define which class of actor or asset the tools will interact with, known as the *Supported Class*, enabling them to be added contextually to the right-click menus.

We are going to explore implementing the Command pattern using transaction nodes in Blueprint to add the ability to undo the process a tool does. To do this, we are going to create a simple tool that rotates the selected objects in the level by 45 degrees.

To begin with, let's create a Utility Blueprint and define its *Supported Class* as Actor (so that we can use the tool on any actor in the world). To do this, follow these steps:

1. Right-click in the content browser and navigate to **Editor Utilities | Editor Utility Blueprint**.

2. From the popup, expand the **All Classes** rollout, select **ActorActionUtility**, and click the **Select** button once it becomes available.

3. Give the new Blueprint an appropriate name such as EU_ActorTransformTools.

4. Open the Blueprint, then on the left side, hover over the **Functions** section of the **My Blueprint** tab. This should reveal the **Override** dropdown; from this, select **Get Supported Class**.

5. This should have automatically opened the **Get Supported Class** function graph. From here, delete the **Parent: Get Supported Class** node and, using the dropdown on the **Return Node** node, select **Actor** as the supported class for any tools built in this Blueprint:

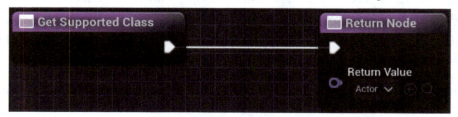

Figure 8.4 – The Get Supported Class function, overridden to set Actor as the supported class

With the Blueprint set up, we can now add our tools to it. We can add tools either as functions or custom events, as mentioned earlier in this section. For this example, we will create our tool as a new function to keep the structure of the tools within the Utility Blueprint tidy:

1. With the Utility Blueprint open, click the + sign on the **Functions** rollout to create a new function; call this function `RotateSelected45`.

2. With the **RotateSelected45** graph open (which should happen automatically) in the **Details** panel, set the **Call in Editor** checkbox to **On**.

With the function set up, let's test that we've enabled the correct things and the function is appearing in the menus when we right-click on something in the viewport, like so:

1. Add some cubes to the level using the **Quickly Add to the Project** button, navigating to **Shapes | Cube**, and dragging one into the viewport.

2. Duplicate the cube a few times by holding *Alt* and dragging the cube using the **Move** tool.

3. Select all of the cubes and right-click on one of them. You should now be able to navigate to **Scripted Actor Actions | Rotate Selected 45**.

At this point, nothing will happen because we have not yet created any logic in the Blueprint function.

Next, we will set up the functionality to rotate the actor (not worrying yet about implementing the Command pattern). The process here is to first identify which actors the user has selected, then use a **For Each Loop** node to rotate the actor in the world. To do this in Blueprint, follow these steps:

1. Drag from the pin on the **Rotate Selected 45** node and add a `Get Selected Actors` node.

2. Drag from the **Return Value** array pin and add a `For Each Loop` node.

3. Connect the output pin from the **Get Selected Actors** node to the **Exec** pin on the **For Each Loop** node.

4. Drag from the **Array Element** pin and add an `Add Actor World Rotation` node.

5. Set the **Delta Rotation** value to **X:**`0`, **Y:**`0`, and **Z:**`45`.

Your function should now look this:

Figure 8.5 – The Rotate Selected 45 function without undo functionality

With the functionality set up, test the function again with the same process as before when we tested that the function was appearing. You should now see that the box rotates.

> **Note**
>
> If you have **Realtime** turned off in the viewport, you will not see the box rotate until you move the viewport for it to update.

With the functionality now working, we want to add the ability to undo the action by implementing the Command pattern using the **Transaction** system. To do this, we need to add three nodes: a **Begin Transaction** node, to start the process of recording the actions, a **Transact Object** node, which is used to identify objects that are about to have a property changed, and an **End Transaction** node, which stops the process of recording actions. Follow these steps:

1. Add a `Begin Transaction` node at the start of the function, between the **Rotate Selected 45** node and the **Get Selected Actors** node.

2. Add a `Transact Object` node as the first part of the for each loop, between the **For Each Loop** node and the **Add Actor World Rotation** node in the **Loop Body** chain.

3. Drag out from the **Completed** pin on the **For Each Loop** node and add an `End Transaction` node, completing the function.

Your function should now look like this; we've added some reroute nodes to the link between the **For Each Loop** node's **Array Element** pin and the **Target** pin on the **Add Actor World Rotation** node:

Figure 8.6 – The completed Rotate Selected 45 function with Transaction nodes

With the function now complete, it is time to test it again. Repeat the steps from before, and the box should still rotate.

The difference now is you should be able to click **Edit** in the top toolbar and see that an option for **Undo Blutility Action** has become available in the **HISTORY** section. If you select it, you should be able to see that any actors that were rotated by the tool return to their previous state:

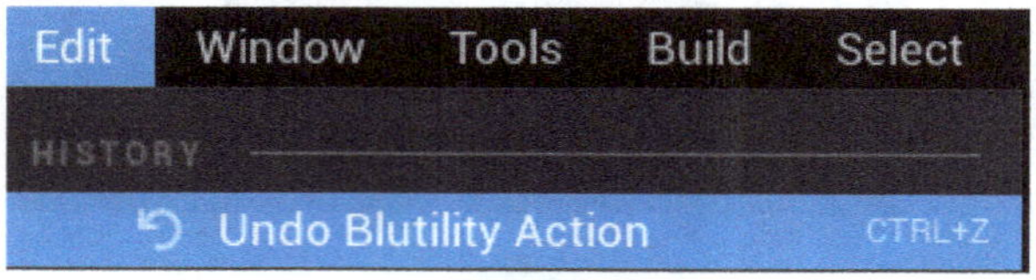

Figure 8.7 – Undo Blutility Action in the HISTORY section of the Edit menu

You can also see **Blutility Action** in the **Transactions** list when you select **Undo History** from the **Edit** menu:

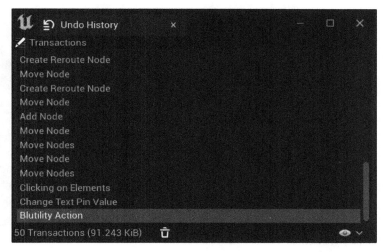

Figure 8.8 – The Transactions list in the Undo History window

With the **Blutility Action** appearing in the undo history and working correctly, the tool is now complete.

When creating tools using Blueprint Utilities, it is considered best practice to always include the ability for the user to undo the effect of the tool, so implementing the Command pattern is key to creating user-friendly solutions.

Command pattern for gameplay in C++

Making our own version of the Command pattern for general use, let's establish a class for our base command. Start by making a child of UObject inside your IDE. We are using UObject so that our class is visible to the engine, but it doesn't need all the extra trappings of AActor, such as a transform. We mark it as Abstract in the UCLASS() macro to make sure no instance will ever be made of this object. This class only needs a virtual execute function as we won't be supporting undo queueing, and initialization will be dealt with differently from standard C++ implementations due to the way Unreal deals with object spawning, let's set that up.

Command.h

```
#pragma once
#include "CoreMinimal.h"
#include "Command.generated.h"


UCLASS(Abstract)
class RTS_AI_API UCommand : public UObject
```

```
{
    GENERATED_BODY()

public:
    virtual void Execute();
};
```

We only need to generate empty function definitions in the body file; there is no need to add any code to the functions as whatever is written there will never run. After creating the base command class, compile everything into the editor layer to make sure there are no errors. With it all compiled, right-click on the `Command` object in the project drawer and create a child C++ class called something like `Command_UnitMove`. This will be the first command linked with our framework. The definition and body are shown next and are simple. The `Init` function is made so that we can set up local variables for the command to execute with, and the `Execute` function is overridden to actually use those values by calling the interfaced function on the reference object. Setup the header file as shown below:

Command_UnitMove.h

```
UCLASS()
class RTS_AI_API UCommand_UnitMove : public UCommand
{
    GENERATED_BODY()

public:
    void Init(AActor* unit, FVector moveLocation);
    virtual void Execute() override;
private:
    TObjectPtr<AActor> _unit;
    FVector _moveLocation;
};
```

As you can see next, the `Init` function actually has no innards. Instead, we use C++ standard constructor overloading syntax to pass the arguments up to the constructors of the internal objects. This is not necessary as code in the body will still work, but it is considered better practice as it uses slightly less memory, but only fractionally. So, let's add the `Init` function:

Command_UnitMove.cpp

```
void UCommand_UnitMove::Init(AActor* unit,
    FVector moveLocation) : _unit(unit)_moveLocation
        (moveLocation)
{}
```

```
void UCommand_UnitMove::Execute()
{
    Super::Execute();
    IControllableUnit::Execute_SetMoveLocation(_unit,_moveLocation);
}
```

The next steps would be to augment the system for consuming input to distinguish when we want to add a command to the queue versus just overwriting what is there. This would be a couple of lines in many different classes, but we have provided the base for you. To inspect what has been made, you can head to the PC_RTS header and body and look at how the input is routed into the pawn via interface calls. The part we are working on next is the character being controlled. There is a function in the AEliteUnit class called QueueMoveLocation_Implementation, which is currently empty. As it stands, our player can right-click anywhere while holding left *shift* for this function to fire. What we would like is for this function to create a new move Command object, initialize it, and store it in a queue. For that, we need a queue, so in the header, add a protected or private TQueue<TObjectPtr<UCommand>> variable, autocompleting the #include instances where necessary, and name it something sensible; we have gone with _commandQueue. The QueueMoveLocation_Implementation function can now check if the character is in the process of moving, and if it is, do exactly what was detailed previously. Create a new command to reify the request, initialize it with values, and add it to the queue:

QueueMoveLocation_Implementation function

```
void AEliteUnit::QueueMoveLocation_Implementation(FVector
targetLocation)
{
    if(!_isMoving)
    {
        _AIController->GetBlackboardComponent()->
            SetValueAsVector("MoveToLocation", targetLocation);
        _isMoving = true;
        return;
    }

    TobjectPtr<Ucommand_UnitMove> moveCommand =
        NewObject<Ucommand_UnitMove>(this);
    moveCommand->Init(this, targetLocation);
    _CommandQueue.Enqueue(moveCommand);
}
```

It is all well and good storing requests to move, but there needs to be a system for clearing the queue as commands are completed, making use of the Dequeue function. This is where the MoveLocationReached_Implementation callback function comes in. This needs to see if there are any commands, and if there are, remove them from the queue and call their Execute function. This is the reason we made sure the command has all its resources injected on creation: so that the `Execute` function can remain parameter-less and therefore useful, as set out in the code below.

MoveLocationReached_Implementation function

```
void AEliteUnit::MoveLocationReached_Implementation()
{
    _isMoving = false;
    if(!_CommandQueue.IsEmpty())
    {
        TObjectPtr<UCommand> command;
        _CommandQueue.Dequeue(command);
        command->Execute();
    }
}
```

With that, the system is functionally complete. There is, of course, no feedback to the user about what currently exists in the queue; that would require a lot more in the way of functions to peek values of commands, but if you wanted to expand this to include different types of move commands or actions, then all you'd need to do is make them from the `UCommand` base, create them somewhere, and add them to the queue.

Our system uses a queue of commands for stacking player commands, but if you wanted an AI to be in control, you could preload it with a set of actions it can perform and let it loose with some form of **Goal-Oriented Action Planning** (**GOAP**). GOAP is an AI design method that abstracts individual interactions with the world from the purpose they fulfill. Stringing these actions together in a chain can create a strategy to achieve a goal. The Command pattern is a great way of implementing this as you can pre-generate a series of commands that make up all the actions an AI can take. Then, the AI strings them together into a queue, as we did previously for the action queue on the controllable unit, to plan its strategy. Executing the commands one by one then hopefully allows the AI to achieve its goal.

By the very nature of making games, there is always some use for the Command pattern, which makes it a good pattern to practice setting as a templated plugin, much as with our next pattern, the state machine, where the Command pattern is used to delay and record logic and the state machine segregates it by use.

Creating the many levels of a state machine

A state machine allows us to separate behavior based on the idea of a *state* – a simple concept, but it has wide-reaching applications from animation state machines to AI logic and contextual player actions.

The simplest of state machines feature two elements: states and transitions. A state machine, at any moment, will either be in a single state or transitioning between two different states, which is why they are a significant part of animation systems, where blending improves the visual quality and feel of characters when receiving input from the player. Let's look at this in more detail:

- A state defines a specific output and/or value of variables. In an animation state machine, the state defines which animation asset should be playing.

- A transition contains the logic that defines when a state machine can transition between two states. Transitions can be adjusted to control the duration as well as utilize curves to further refine the blend weighting between the two states throughout the duration of the transition.

Figure 8.9 shows a simple state machine with three states and four transitions. Each transition has a start state and an end state. Transitions can exist in either direction, as shown in the example, with there being two transitions between **State 1** and **State 3**, one from **State 3** to **State 1**, and another from **State 1** to **State 3**:

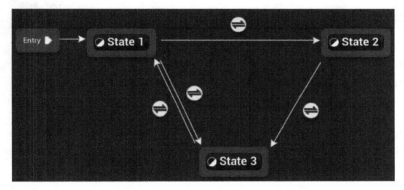

Figure 8.9 – A state machine with three states and four transitions

Exploring animation state machines

A simple character animation state machine might look like the example provided in *Figure 8.10*. This shows a character animation state machine that can stand still (idle), move (denoted by the run state), and jump (which uses three states).

A good example to understand the use of transitions is the setup of a three-part jump animation system:

- The transition from **Idle** or **Run** into **JumpStart** is controlled by the player input (when the **Jump** button is pressed) or, more commonly, with physics (when the character is in the air)

- The second transition into **JumpLoop** occurs when the **JumpStart** animation is near completion, either controlled via logic within the transition or using the automatic option, which starts the transition purely based on the duration of the transition

- The third transition to **JumpEnd** (typically a landing animation) occurs when the character actor returns to contact with the floor – that is, when the character is *NOT* in the air

- The final transitions, back to **Idle** or back to **Run**, are, then, similar to the second transition, based on the remaining time of the landing animation in the **JumpEnd** state:

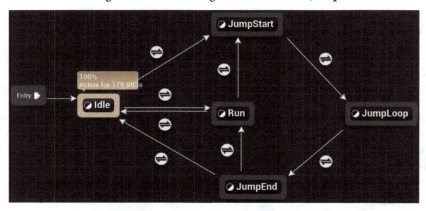

Figure 8.10 – An animation state machine for a character

A single state can be connected to any number of other states via the use of lots of transitions in a one-to-many relationship; however, a state machine can be refined with the use of a conduit node that allows for a one-to-many, many-to-one, or many-to-many relationship within a single node. Conduits contain no state information (such as an animation); they simply act, as their name suggests, as a conduit between states, simplifying the need for an excessive number of transition lines between states to achieve the same many-to-many relationship, as illustrated here:

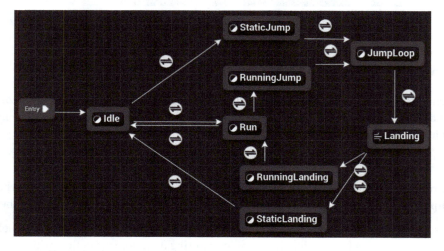

Figure 8.11 – A more complex state machine including a conduit node

The system shown in *Figure 8.11* includes a system to select different jumping and landing animations for when the character is standing still (idle) or when it is moving. A conduit is used to decide which of the two landing animations should be selected:

- The transition out of **JumpLoop** into the landing conduit remains the same as the previous system, based on when the character is no longer in the air
- The transitions out of the conduit into **StaticLanding** or **RunningLanding** are based on the velocity of the character where if the velocity is 0, the **StaticLanding** state will be selected, and if the velocity is greater than 0, the **RunningLanding** state will be selected

While this, at this point, doesn't offer much more efficiency for the number of transitions, it does allow the separation of checks and simplifies the logic used to identify which transition to use. This makes the system much easier to expand, for example, by adding a different landing if the character has fallen a larger distance or if the character is traveling backward.

Animation state machines can become very complex and even include state machines within state machines, where a state utilizes its own state machine to determine its own output. The complexity of the system depends upon the required solution and how many different animations need to be considered.

We can create many styles of state machines that make them useful for different purposes. We'll be starting the journey by looking at the simplest form: an enum/switch implementation where the state is determined by a simple variable.

Enum/switch implementation

If all you need to do is separate mutually exclusive logic, then an enum-style state machine is all you require. Creating an enum with a value for each state is the start. Unreal requires enum definitions to have specific tags to be usable in Blueprint, where you may want to visualize things to make debugging easier. The important thing is the UENUM() block. In the following code snippet, we can see that it includes the BlueprintableType property. This allows the editor layer to both serialize C++ variables of this type into details panels and also create Blueprint-level variables of this type. The enum is also defined as a class inheriting from an unsigned integer. This is different from standard C++ where an enum is its own type. Unreal naming convention would have you prefix your enum type name with E, but this is not a requirement for it to function. You will have to decide what size unsigned integer to inherit from. Here, we have shown a uint8 type, which would give you 8 bits of size, meaning that you can either have 256 values as standard or 8 values if using the enum to pack Boolean values as flags. In most cases, uint8 will provide the necessary space. Each value also has a UMETA() block that gives the flexibility of a different display name in the editor:

UENUM example definition

```
UENUM(BlueprintType)
enum class EState : uint8
```

```
{
    State1      UMETA(DisplayName = "First State"),
    State2      UMETA(DisplayName = "Second State"),
    State3      UMETA(DisplayName = "Third State"),
};
```

Making an enum variable to track our current state allows us to segregate our logic into different case blocks of a `switch` statement. While an integer variable could be used for this same job, the enum only has a small additional cost and makes our code more human-readable. Anywhere we need to execute different logic based on the state, we can use a `switch` case, as in the following example. This is in place of many Boolean variables combined in complex amounts of `if` statements, instead setting it out like this:

Example state machine switch statement

```
Estate _State;

void SomeFunction()
// some code…
switch(_State)
{
case Estate::State1:
    //State1 code
    break;
case Estate::State2:
    //State2 code
    break;
case Estate::State3:
    //State3 code
    break;
default:
    //unhandled state code
    break;
}
```

> **Helpful tip**
>
> Save yourself time by leveraging your IDE's code autocomplete functionality to create the template `switch` statement followed by all the `case` statements with a few keyboard presses. In Visual Studio, this requires autocompleting the statement. Without clicking away, press *Enter* twice after adding the variable to the expression brackets; this should engage the autocomplete and paste in the template structure. Rider uses the *Alt + Enter* autocomplete tool to do this.

The enum/switch method works when all you need to segregate is logic. As soon as variables that only have use in some states get involved, we can enforce the **single responsibility principle (SRP)** from *Chapter 1* and bundle the full state, variables, and logic into a class. This helps with readability and expansion once our state machine starts to grow. These new states are called *stateful* states because they have some essence of memory attached to them, as opposed to the enum/switch method, where states are pure logic and therefore stateless.

Static versus instanced states

When using stateful states, there is a decision to make. We can define our states as either static or instanced. Static states only exist once in stack memory, helping with overall memory size if multiple machines have reference to the logic. This can be especially helpful for AI state machines being used by large numbers of actors. Instanced states must be created and likely exist in heap memory. Instanced states are necessary when the state of a state – that is to say, the current value of the variables it holds – matters to the actor running the state machine. Something such as a heavy attack charge level cares about which state machine it is running on and therefore must be in an instanced state; otherwise, all state machines running that state will share a charge level.

To implement states as separate classes, there must be a parent state class that has the basic functionality of a state. The following code shows instanced states, but a static state would be largely the same with the static keyword. In this example, we are using the Update() function to return an EState value that will inform the machine running this state when to change state and which one to change to. It also has other benefits of using separate classes for states, enter, and exit logic. Usage of this can vary wildly, but it gives gameplay programmers a hook for the moment a state starts and finishes to run extra logic. Lastly, marking all the functions as purely virtual will make the class abstract in nature and ensure all child states have these functions implemented in some capacity:

Example state base class header

```
class State
{
public:
    virtual EState Update() = 0;
    virtual void Enter() = 0;
    virtual void Exit() = 0;
}
```

The actual machine is very simple; it consists of a `State` variable to hold the current state being run and some kind of collection for the rest of the states to be created into and held in. This is where a static implementation may differ; states are still held as references, but they don't need to be instantiated. Setup the header for the instanced state machine as shown below:

Instanced state machine implementation header

```
class SomeClass
{
public:
    SomeClass();
    void Update();

private:
    State* _State;
    EState _CurrentState;
    Map<Estate, State*> _StateLibrary;

    void ChangeState(EState nextState);
};
```

This code makes use of a `Map<>` collection. This doesn't strictly exist in base C++ but it is analogous to Unreal's `TMap<>` collection, which stores key-value pairs and indexes by key. The body therefore can be setup like this:

Instanced state machine implementation body

```
SomeClass::SomeClass()
{
    State* tempState = new ExampleState();
    _StateLibrary.Add(EState::example, tempState);
    tempState = new OtherState();
    _StateLibrary.Add(EState::other, tempState);
    // Do this for each state the state machine requires
    _CurrentState = EState::example;
    _State = _StateLibrary[_CurrentState];
    _State->Enter();
}

void SomeClass::Update()
{
    EState next = _State.Update();
    if (next != _CurrentState)
```

```
    {
        ChangeState(next);
    }
}

void SomeClass::ChangeState(EState nextState)
{
    _State->Exit();
    _CurrentState = nextState;
    _State = _StateLibrary[_CurrentState];
    _State->Enter();
}
```

In Unreal, making objects of base C++ classes can make your code faster as you are ignoring all the overhead of Unreal's Editor layer, but if you want to make an instanced state machine that can be balanced by designers, it is a good idea to create your states as Actor components. That way, you could build the entire system to work through the editor using custom slate tools.

The next bottleneck in the system comes into play when multiple unconnected factors influence different parts of the behavior. This could mean the state machine is governing locomotion and world interaction. With the current system, a new state would be needed for each combination of possibilities, such as a running heavy attack and a crouched heavy attack. The heavy attack logic is likely to be duplicated between these states, which should be setting off alarm bells. There is a solution in concurrent state machines.

Concurrent state machines

The term *concurrent* simply means to be running at the same time as something else. It is a term thrown about a lot in the networking sphere, but here, it means something far simpler. When there are two or more areas of control that never cross, we can create a state machine for each area and run them at the same time. Expanding our implementation from the preceding section, the following code shows that each state machine exists and updates side by side. The rest of the setup will also need to be doubled. Only minor changes are needed to the ChangeState() function to accommodate the upgrade, making it machine-agnostic. The State*& argument here is used to pass the pointer by reference instead of the value being pointed to, as it is the pointer we need to retarget. This can be seen below.

Concurrent state machine implementation

```
void SomeClass::Update()
{
    EMoveState nextMove = _StateMove.Update();
    EAttackState nextAttack = _StateAttack.Update();
```

```
    if (nextMove != _MoveStateTracker)
    {
        ChangeState(nextMove, _MoveStateTracker, _MoveState);
    }

    if (nextAttack != _AttackStateTracker)
    {
        ChangeState(nextAttack, _AttackStateTracker, _AttackState);
    }
}

void SomeClass::ChangeState(EState nextState,
    EState& stateTracker, State*& stateMachine)
{
    stateMachine->Exit();
    stateTracker = nextState;
    stateMachine = _StateLibrary[stateTracker];
    stateMachine->Enter();
}
```

The next area for improvement will be noticed most in state machines that govern variant behavior where multiple states might share some base functionality but differ slightly. An example of this could be that a crouching state has different state transitions from a running state but the logic for executing movement is the same. In this case, we can take advantage of the fact that all states inherit from a parent and add some more layers with a hierarchical state machine.

Hierarchical state machines

As the name suggests, the state inheritance tree being pulled into a larger inheritance hierarchy with base behavior for groups of states creates a hierarchical state machine. Everything within this actually functions exactly the same as the base state machine; you just need to make sure that the implementations of each function call the base version. This is more of a code architecture principle to reduce memory footprint and speed up development. *Figure 8.12* shows an example state inheritance tree for character locomotion and how we can use the grounded mid-level state to create several other states that implement grounded movement logic without needing to write it out a bunch of times. This improves maintainability as well since we are centralizing behavior:

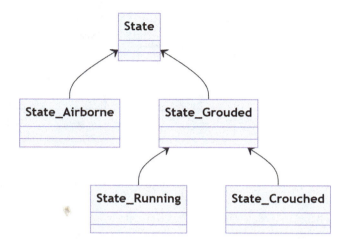

Figure 8.12 – UML diagram showing an example inheritance structure for a set of hierarchical states

The final step is not to do with the execution logic but the transitions. As it currently stands, each state transition must know which state it's heading to. This can work for simple state machines, but as soon as you have an interrupt action that can be accessed from anywhere and must return execution to where it came from when finished, we have a problem. Storing the information on where the state was entered from doesn't really solve the problem, because what if the state we were in had a state that must be preserved? The answer is the very fancy pushdown automata level of a state machine.

Pushdown automata

As just mentioned, pushdown automata aim to provide a way to enter states with a breadcrumb trail to follow back out if the need arises. We achieve this by storing the current path of states in a stack. The only code change here is that the tracker and state pointer variables get merged into a struct and stored in a single state stack instead. Execution is simply run the same way on whichever state happens to be on the top.

This allows us to push interrupt states from anywhere in the machine, then pop that state off and continue from where we were – another simple idea that provides a lot of utility. Prime example usage would be a state machine that governs animations on a character that can move and attack. Attacks take over the animation solver until they are finished. The character should then return to whichever state they were in, be that crouched or standing idle. *Figure 8.13* shows how the stack changes as the interaction progresses:

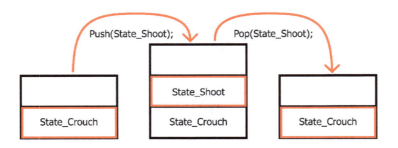

Figure 8.13 – Storyboard of a pushdown automata state stack over the course of an attack input

The state machine, as we have covered, has many forms. The real silver bullet with this pattern is realizing that none of these forms is mutually exclusive. You can make a hierarchical pushdown automata state machine that runs concurrently with another. The flip side of that is noticing that if you are using a fully featured hierarchical pushdown automata state machine in earnest, there is probably a simpler pattern that solves the same problem in a much more elegant way.

Summary

Altogether, we have covered three patterns that you can build into any future project: the Singleton pattern, which makes sure there is – and only ever will be – one instance of an object that exists, the Command pattern, which provides the utility of separation between the request for an action and that action happening, and finally, the State pattern, which separates our mutually exclusive logic.

At this point, you should understand that Singleton does work in some cases, but it has its drawbacks. The Command pattern can be used for so many different things that you should probably make a version of it as a template library for use in all future projects, and the State pattern has so many layers that it can ruin its own usefulness with depth.

The next chapter will explore some behavioral patterns that solidify the behavior of a class in different ways to improve the expandability of our systems – namely, the type object pattern, which we would posit as the most useful pattern in game development for content creation.

9

Structuring Code with Behavioral Patterns – Template, Subclass Sandbox, and Type Object

This chapter focuses on the three most common structural patterns. Structural patterns allow us to plan our code with the end usage in mind. For example, if we know that the end users of our system are likely to be designers with no code experience, we could plan to use the type object pattern to provide a system for easy dynamic expansion. We have already covered some of the concepts around code structure when we discussed using interfaces and events to achieve anonymous modular design in *Chapter 7*. The three patterns in this chapter (template, subclass sandbox, and type object) are a little more zoomed-in in terms of scope compared to what we have looked at before. The first two are interchangeable depending on your preference, both working as extensions to the standard inheritance property of the C++ programming language. The last is by far the most useful pattern in game design, giving designers the ability to define variants of classes with ease without getting in the programmers' way while they make new functionality.

The topics we will cover in this chapter are the following:

- Exploring the **Template** pattern
- Understanding **Subclass Sandbox**
- The **Type Object** pattern

Technical requirements

Unlike previous chapters, we will be starting with the project files in the `chapter9` branch on GitHub, which can be downloaded from `https://github.com/PacktPublishing/Game-Development-Patterns-with-Unreal-Engine-5/tree/main/Chapter09`

We have made a few small changes to the project to facilitate the following pattern examples. These changes are a bit too tedious to walk through in this chapter but if you would like to see what has changed, then download both this and the previous chapters' branches and run them through a diffing program such as DiffMerge.

Let's get started.

Exploring the template pattern

The template pattern exists as an extension to standard inheritance, where we define a structure in an abstract parent class and the children are given the opportunity to override the pieces of that structure. They can change how individual parts function but not the order of execution. The simplest example of this within Unreal is the `AActor` base class. Any child of AActor gets access to the Begin Play, Tick, and End Play events, to name but a few. The child class can hook functionality onto these events, and they will fire when expected. The constraint we place on inheritance to make this into the template pattern is that the child has no way of changing the order or timing of these events. There is no way to make End Play fire before Begin Play as this order has been defined in the parent.

We can see an example of a class implementing the template pattern in the following code. The `ProcessGame()` function is the only one with a body, defining the order of the private abstract function:

Template pattern parent pseudocode

```
public class AActor
{
public:
void ProcessGame() {
BeginPlay();
while(gameRunning)
    Tick();
    EndPlay();
    }
protected:
    abstract void BeginPlay();
    abstract void Tick();
    abstract void EndPlay();
}
```

We can see the `AActor` class has four functions with only one of them not marked as abstract or virtual. This function is our spine that defines the order of execution. From there, each of the other functions may have some form of implementation in the base but they are designed to be ultimately overridden. The best situation is for each of these extra utility functions to be abstract with no implementation in the base to retain the lightweight nature of the pattern, but if that means that a lot of the implementations will be repeated, then a virtual base with the common code would be better. The goal is to keep this class as light as possible while cutting down as much repetition as we can.

> **Important caveat**
>
> Strictly speaking, the actual structure of `AActor` doesn't implement the Template pattern, as shown in the pseudocode; we collapsed the tree a little to make a point. In reality, that loop is dealt with by the world object and filtered down to all the actors and subsystems within it.

To demonstrate this, we are going to give our elite unit some weapons. In the GitHub branch for this chapter, the `EliteUnit` C++ class has already been augmented to spawn a weapon from a base class defined in the class defaults. This spawned weapon is an actor that we have attached to a `SceneComponent`, positioned just in front of the character. The code is also modified to call the fire function on this weapon instead of running a line trace from within the character. As it currently stands, this code will not run as there are no classes available for the character to spawn from that are not marked as abstract. Our first step is to build the template parent class as a child of this generic weapon in C++. We can then leave the creation of the functional child weapons to a Blueprint, giving us an excellent way to understand how the two systems work together and allowing programmers to do the groundwork in C++ before the technical designers and designers explore variations in Blueprints. This allows for quick prototyping and iteration of weapon designs to achieve the intended game feel without breaking any of the underlying systems.

Building the template (parent class)

As mentioned earlier, our first step is to make a new C++ child class of `AWeapon_Base`. With that made, we can set about putting the template main function in place which, as we can see in the code below, is called `Fire()`. The parent class has a lot of the data and a core system for dealing with a brief cooldown delay between shots. These variables have been added at the top level as they will be universal to all weapon types and so it makes sense to consolidate them in the common parent. This does mean that the `Fire()` function here is an override, whereas in a straight implementation of the template pattern, this would be at the top level. We are only doing things differently here to create a shared hierarchy with the Sandbox pattern that we'll see in the next section to highlight the differences.

The other functions in the template weapon base class include a public function for reloading, as you may want to actively call that function even though we will be triggering it by default anyway. This is followed by a bunch of utility functions that are marked as either `BlueprintNativeEvent` if they have a default implementation in the base, or `BlueprintImplementableEvent` if they do not. Each of these protected functions exists to be overridden by the children to change the behavior of the weapon:

TemplateWeapon_Base.h excerpt

```cpp
UCLASS(Abstract)
class RTS_AI_API ATemplateWeapon_Base : public AWeapon_Base
{
    GENERATED_BODY()

public:
    virtual void Fire() override;
    UFUNCTION(BlueprintCallable, BlueprintNativeEvent)
    void Reload();

protected:

    UFUNCTION(BlueprintNativeEvent)
    bool CheckAmmo();
    UFUNCTION(BlueprintImplementableEvent)
    void ProcessFiring();
    UFUNCTION(BlueprintImplementableEvent)
    void PlayEffects();
    UFUNCTION(BlueprintNativeEvent)
    void UpdateAmmo();

};
```

The only function in the implementation of this class is the template spine we are calling `Fire()`. In this function, we define the execution order of all the other functions in a way that becomes concrete. In the following implementation, we check whether we can fire; if not, we use an early return to exit the function. Then we check whether we have the ammo to fire; if not, then we'll reload. If the weapon is able to fire, then the process outlined by this function starts with the effects, then runs the gameplay logic for actually firing with the reduction in ammo calculated after that. The last thing it does is call the parent, where we have set up the code to deal with the firing delay (which is used to prevent the fire button being spammed or exploited with auto-clickers and macros):

TemplateWeapon_Base.cpp main template function

```cpp
void ATemplateWeapon_Base::Fire()
{
    if(!_CanFire) return;

    if(CheckAmmo())
```

```
    {
        PlayEffects();
        ProcessFiring();
        UpdateAmmo();
        Super::Fire();
        return;
    }

    Reload();
}
```

The next few functions are just default implementations that make sense and are likely to be repeated code if not defined here. The reload function resets the ammo counter and calls the fire delay reset function. CheckAmmo is a one-line Boolean check that could be made pure as an extension to this. Finally, UpdateAmmo just takes the predefined _AmmoPerFire from our ammo counter:

TemplateWeapon_Base.cpp utility functions

```
void ATemplateWeapon_Base::Reload_Implementation()
{
    _CurrentAmmo = _MaxAmmo;
    Handle_FireDelay();
}

bool ATemplateWeapon_Base::CheckAmmo_Implementation()
{
    return _AmmoPerFire <= _CurrentAmmo;
}

void ATemplateWeapon_Base::UpdateAmmo_Implementation()
{
    _CurrentAmmo -= _AmmoPerFire;
}
```

That forms the C++ base for our template pattern. All that is left is to make an implementation or two with Blueprint child classes in the editor.

Creating child classes

We are now going to create two child classes in a Blueprint. This task offers us a great opportunity to explore how C++ and Blueprints can work together to create efficient solutions. To do this, we need to create new child Blueprints, one for a pistol and a second for a shotgun. These weapons both utilize a line-trace approach with variation for multiple projectiles in a spread. This allows us to focus on creating two different solutions with a small but significant change to build a solid example for the template pattern. To do this, we are going to use the same menu from before, but instead of selecting **Actor** from the dialog, we need to dig a bit deeper to find the C++ parent class:

1. Start by right-clicking in the Content Browser and clicking **Blueprint Class**.

2. If it isn't already visible, expand the **ALL CLASSES** rollout by clicking the title.

3. In the search box, type `template` to reduce the number of options displayed in the results box, as shown in *Figure 9.1*.

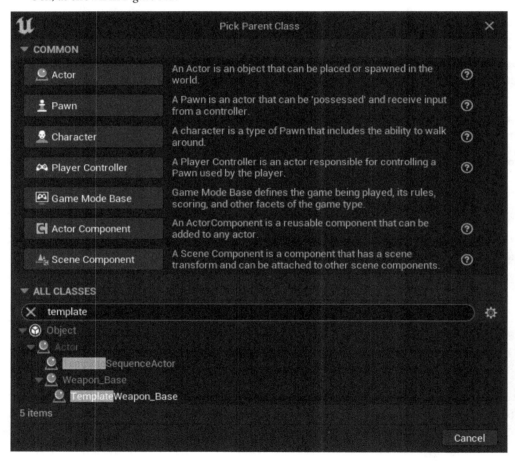

Figure 9.1: The Blueprint creation window for a template weapon

4. Select **TemplateWeapon_Base** from the list, then click **Select**, which will only appear once you've made your selection.

5. Name this new Blueprint `BP_TemplatePistol`.

With the pistol child Blueprint created, we next need to create a child Blueprint for the shotgun. To do this, repeat the preceding steps but name the second blueprint `BP_TemplateShotgun`.

Now that the child classes have been created, we can move on to creating the behaviors inside them, making use of the template pattern to only modify the part of the class necessary to achieve the desired functionality of each child.

Template pistol

Now that we have our Blueprint for the Pistol set up, we can move on to overriding the parts of the template that will be unique to this class. The `CheckAmmo` and `UpdateAmmo` functions will remain the same for our pistol class as in the parent template. We will be overriding `PlayEffects`, `ProcessFiring`, and `Reload`.

We've overridden functions before, namely in *Chapter 8* when looking at utility blueprints. The process here is exactly the same.

> **Note on overriding C++ functions**
>
> When we override functions in a Blueprint that are included within a C++ parent, we don't get a choice between creating a function or an event. The result we get when selecting to override the C++ function will depend on whether the function has a return or not. If a function does not contain a return, the override in the Blueprint will automatically create an Event node in the main Blueprint Event Graph. If the function does have a return (like `CheckAmmo` does), the override will provide a Blueprint function.

First, we are going to override the `PlayEffects` function. This is where we would typically add a muzzle flash particle effect and play an animation and a sound. In the interest of keeping the repository for this exercise small, and the number of steps short to allow us to focus on learning, we will just add a sound effect. Again, in the interest of keeping the required download small, we will use a sound from the Engine Content as opposed to finding a purpose-selected sound for the gun. Let's do this as follows:

1. Open the Blueprint and on the left side, hover over the **Functions** section of the **My Blueprint** tab. This should reveal the **Override** dropdown. From this, select **Play Effects**.

2. This should have added an **Event Play Effects** node to the Event Graph.

3. Drag from the pin on the **Event Play Effects** node and add a `Play Sound at Location` node.

4. Expand the **Sound** dropdown and search for `VR_Teleport`. If nothing appears, you will need to enable **Show Engine Content** from the **Settings** dropdown, which can be accessed by clicking the cog icon in the top right of the dropdown.

5. Drag from the **Location** pin to the **Play Sound at Location** node and add a `Get Actor Location` node.

Figure 9.2: The overridden Event Play Effects function in the Blueprint

Next, we will override **ProcessFiring**. This is where we are going to put all of our firing logic. The precise contents of this function will differ depending on the type of weapon. For the pistol, we can use a simple line trace similar to the approach taken for attacking enemies that we created in *Chapter 4* when making the Behavior Tree for the Elite Unit. To begin, let's set up the trace:

1. Drag from the pin on the **Event Process Firing** node and add a Line Trace by Channel node to the graph.

2. Drag in the **Fire Point** component from the **Components** tab. From the resulting node, drag out and add a Get World Location node and then repeat this again to add a Get Forward Vector node.

3. Connect the **Return Value** pin on **Get World Location** node to the **Start** pin on the **Line Trace By Channel** node.

4. Drag from the **Return Value** pin of the **Get Forward Vector** node and a Multiply (x) node by typing * in the search box.

5. Right-click the lower pin on the **Multiply** node and select **Convert Pin | Float (single-precision)**. This will turn the yellow pin green, indicating that the variable expected is no longer a vector but now a float.

6. Drag from the green pin and add a Get Range node. This will add the Range variable defined in the parent class.

7. Once again drag from the **Get World Location** node and create an **Add** node (reached by typing +).

8. Connect the lower input vector pin of the **Add** node to the output vector pin of the **Multiply** node.

9. Connect the output of the **Add** node to the **End** pin on the **Line Trace By Channel** node.

Next, we need to tell the line trace to ignore the character when shooting, or else the only character the gun will damage will be the one firing it:

1.   Right-click in the event graph and create a `Get Owner` node.

2.   Drag from the **Return Value** output of the **Get Owner** node and add a **Make Array** node.

3.   Connect the **Array** output pin of the **Make Array** node to the **Actors to Ignore** pin on the **Line Trace By Channel** node.

The line trace is now set up and should look like the graph shown in *Figure 9.3*.

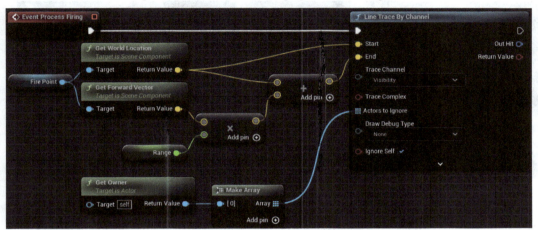

Figure 9.3: The Line Trace node with the inputs setup

> **Drag and drop variables defined in the parent class**
>
> If you'd like to be able to drag the variables we defined in the parent class into the graph in the same way you would with a variable created in a blueprint, click the cog in the top right of the **My Blueprint** tab and enable **Show Inherited Variables**.

Next, we need to do something with the result of the trace. In this case, we are going to use the Apply Damage approach, just as we did in *Chapter 4*:

1.   Drag from the **Out Hit** pin on the **Line Trace By Channel** node and add a `Break Hit Result` node.

2.   Drag from the **Hit Actor** pin on the **Break Hit Result** node and add an `Apply Damage` node.

3. To make the graph easier to read, click the collapse arrow (^) on the bottom of the **Break Hit Result** node. This will collapse the node to only show the top two bool pins and any pins that have been used, hiding any unused pins. In this case, just the **Hit Actor** pin will remain visible along with the top two bool pins (**Blocking Hit** and **Initial Overlap**).

4. Drag back from the **Base Damage** pin and add a `Get Damage Per Hit` node, once again grabbing a variable defined in the parent.

5. Lastly, drag from the **Return Value** bool pin on the **Line Trace By Channel** and add a `Branch` node (`If`) and connect the three nodes together as shown in *Figure 9.4*.

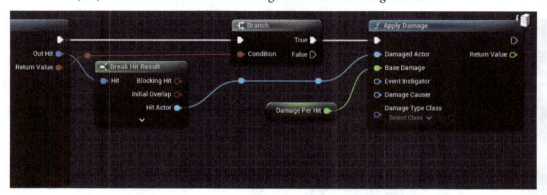

Figure 9.4: The result of the trace used to apply damage to the enemy unit

The last thing to do is to check the values of each of the inherited variables. You can do this by clicking on the variable nodes or, if you chose to show inherited variables, you can select them from the variables list in the **My Blueprint** tab.

Ensure that the **Default Value** for **Range** is `1000.0` and **Damage Per Hit** is set to `20.0`.

With the firing of the weapon sorted, we can next turn our attention to the reload function. Unlike the previous two overrides, the reload function is not an abstract function. The parent template class already handles the functional aspect of the reload, adding the ammo back into the weapon, so we need to ensure we retain that in the child class. We can then add all the extra elements such as sounds, animations, particle effects, and so on. Once again, let's keep this simple by just adding a sound. First, let's override the Reload function and ensure we are maintaining the functionality from the parent class:

1. Override the **Reload** in function in the same way as we did for the previous two functions. This will add the **Event Reload** node to the **Event Graph**.

2. Right-click on the **Event Reload** node and select **Add call to parent function**. This will add a **Parent: Reload** node for you.

3. Move the **Parent: Reload** node beside the **Event Reload** node and connect them together.

Figure 9.5: The Event Reload and Parent: Reload nodes

The **Parent: Reload** node tells Unreal to do all the steps from the parent version of this function at this point in the logic chain.

Now that we have the behavior from the parent included in the child, we can add the sound. We do this just in the same way as we did for the **PlayEffects** event, however this time, we will select the **Gizmo_Handle_Clicked** sound.

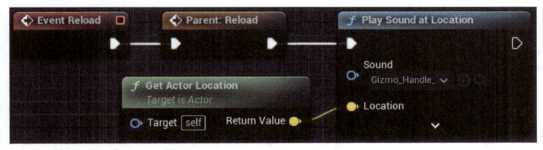

Figure 9.6: The completed Event Reload function in our Blueprint

With the reload function sorted, our pistol is complete. Let's test it out on the character to make sure it works. The **EliteUnit** modifications included an automated system for spawning and attaching weapons to the character based on a drop-down box. Set the weapon as follows:

1. Select the **BP_EliteUnit** that is in the level.

2. Search in the details panel for Weapon to Spawn.

3. Change the value in the drop-down box to **BP_TemplatePistol**.

If you see **TemplateWeapon_Base** in your drop-down list, then you need to make sure the UCLASS block includes the Abstract property so that no one can ever spawn an instance of the base class.

With the pistol now set up, tested, and functional, let's look at the shotgun.

Template shotgun

For the sake of simplicity and to avoid this being a very long chapter, with the shotgun the only unique approach we will apply is to the **Process Firing** function. Because the Play Effects and Reload functions are abstract in the parent, we must override them, so before we get started on the Process Firing for the shotgun, replicate the other two functions from the pistol in the **BP_TemplateShotgun** Event Graph. You can do this by creating them again manually or by copying and pasting them between graphs.

With that done, let's move on to sorting the firing out. Rather than going through this version step by step, let's look at the differences so you can make the necessary changes to create a shotgun rather than a pistol. Begin by copying over the firing logic from the pistol before modifying it to replicate each of the changes that follow.

Unlike a pistol, we need to consider that when fired, a shotgun doesn't fire a single projectile. Rather, shotgun shells often contain between nine and eighteen small pellets that are fired out of the barrel in a cluster.

To replicate this, we simply increase the number of traces we perform and add variation to the direction, maintaining the same start point (the end of the barrel) but adjusting the end point of each trace to mimic the spread of the shot exiting the barrel

Let's get started by using multiple traces to replicate the shotgun shell's behavior.

Adding a for loop to fire multiple pellets

To perform multiple line traces, we need to add a For Loop as the first node in the **Event Process Firing** logic chain, connecting the **Line Trace By Channel**, **Branch** and **Apply Damage** nodes into the **Loop Body**.

The **For Loop** node will complete each of the steps in the *Loop Body*, incrementing its index from the **First Index** value to the **Last Index** value.

For the shotgun, set **First Index** to 1. Typically, we would use 0, however we are going to create an integer variable called NumberOfPellets for the **Last Index** and set the **Default Value** of **NumberOfPellets** to 5. If we were to keep **First Index** at zero, the shotgun would fire six pellets, so instead of requiring a designer to remember to reduce the variable by 1 to get their desired number of traces, we simply start at 1 to make the system more user friendly.

Figure 9.7: The For Loop node and Number Of Pellets variable

With multiple traces firing, we now need to make each trace slightly different so that we aren't just doing multiple identical traces. Let's do that next.

Adding spread to the pellet trajectories

To add variation to the **End** vector input of the **Line Trace By Channel** node, we need to rotate the result of the **Get Forward Vector** node slightly before multiplying it by the range.

To do this, we add a `Random Unit Vector in Cone in Degrees` node. This uses a **Cone Dir** input (the initial direction the cone is facing) and a **Cone Half Angle in Degrees** input (the amount of rotation to be applied from the center line of the cone).

For **Cone Dir**, we connect in the **Return Value** of the **Get Forward Vector** node.

For **Cone Half Angle in Degrees**, create a new **Float** variable (either by promoting the pin to a variable or from the **Variables** rollout) and name it `HalfSpreadDegrees`.

Set the **Default Value** of **HalfSpreadDegrees** to `15`.

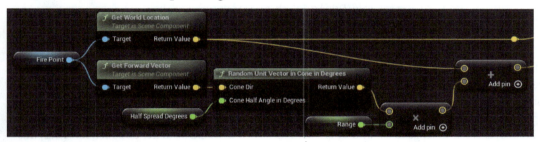

Figure 9.8: The adjusted firing trajectories

Once you've added the **Random Unit Vector in Cone in Degrees** node, the calculations should be connected together as shown in *Figure 9.8* with the vector connections going into the **Line Trace By Channel** node.

With all that now set up, you are ready to test the shotgun. Change the variable once more on the Elite Unit in the level and give it a go.

Once that's all finished and working, let's move on to another parent-child structural pattern where everything is the exact opposite.

Understanding subclass sandbox

The Subclass Sandbox pattern takes the idea of the template's limited extension through subclasses (to provide security) and applies it the exact opposite way round. Here, the children define the order of execution for a set of pre-defined code blocks through an abstract spine function. These blocks take the form of functions that are defined in the parent class and can never be overridden. Each function deals with one thing to do with an external system in a standardized way. The following pseudocode makes a better visual point of how this is literally the opposite of the template pattern we explored previously, where everything previously marked as abstract gets functionality and the one function we had code in is now abstract:

Subclass sandbox pattern parent pseudocode

```
public class Sandbox_Parent
{
public:
    abstract void DoAThing();
protected:
    void PlaySound() { //Plays sound correctly }
    void FireParticle() { //Fires particle correctly }
    void AddForce() { //Adds force correctly }
    void DealDamage() { //Deals damage correctly }

}
```

The workflow for this pattern starts with the programmer building black-box tools for the technical designer to string together in interesting ways. This can lead to interesting innovation as the programmer isn't making tools to specification, rather just for the sake of having tools, which leaves their application open for the technical designer to interpret. The flip side of this is that there could be a lot of wastage where tools either don't get used or they are used in ways they could be better designed for. Thus, this process works better as an iteration on both sides to ensure that the toolset is useful and efficient, as well as actually being used.

Enough speaking in the abstract: let's look at the implementation of this for our weapons alongside the template pattern.

Building the sandbox (parent class)

The first step is to copy what we did in the template section and make another C++ child of our Weapon_Base class, but this time call it SandboxWeapon_Base. The whole point of this pattern is that the Blueprint children will implement the Fire function in whatever way they see fit, but for that we would have to add UFUNCTION(BlueprintNativeEvent) at the very least. However, because we have joined our patterns together with this common parent, it means the function has to be marked as virtual and so cannot have a standard function specifier. This will never be the case in any production code. It is only a problem here due to us showing both patterns linked by a common parent. Our solution for this is a second Fire function called SandboxFire. We will pass the execution off to this function within Fire so that everything behaves as normal. No sensible project architecture would implement both patterns side by side like this, so it should not be an issue in your future projects.

The other thing we must define are the building block functions, which we have chosen to keep similar to the previous example, but for which you can make as many as you deem necessary. The key is to keep each function short and to the point as their purpose is to standardize the method for interacting with external systems so that future changes are easily maintained. The most important aspect of the functions laid out in the following code are that they are protected and marked as `BlueprintCallable`. This means they are just for children of this class to use, not override, and also ensures they have no external access.

Lets start with the base class for the sandbox weapon:

SandboxWeapon_Base.h excerpt

```
UCLASS(Abstract)
class RTS_AI_API ASandboxWeapon_Base : public AWeapon_Base
{
    GENERATED_BODY()
public:
    virtual void Fire() override;

    UFUNCTION(BlueprintImplementableEvent)
    void SandboxFire();
    UFUNCTION(BlueprintCallable)
    void Reload();

protected:
    UFUNCTION(BlueprintCallable)
    bool CheckAmmo();
    UFUNCTION(BlueprintCallable)
    void LinetraceOneShot(FVector direction);
    UFUNCTION(BlueprintCallable)
    void PlaySound(USoundBase* sound);
    UFUNCTION(BlueprintCallable)
    void UpdateAmmo();
};
```

With the header declared, we can turn to the definitions in the following implementation. Our `Fire` function override simply calls `SandboxFire()` to pass the signal through to a function with the correct properties. As `SandboxFire` is marked as `BlueprintImplementableEvent` it doesn't have a definition in this class and instead can be completely left to the Blueprint children to define. Our `Reload`, `CheckAmmo`, and `UpdateAmmo` functions are basically the same as before, but this is dependent on the systems that are being hooked into and where it is best to build lots of small specific functions. `PlaySound` is a more specific version of `PlayEffects` from before, where we were able to leave the implementation to the designer, but now we must be specific in function name and use:

SandboxWeapon_Base.cpp simple function definitions

```cpp
void ASandboxWeapon_Base::Fire()
{
    SandboxFire();
}

void ASandboxWeapon_Base::Reload()
{
    _CurrentAmmo = _MaxAmmo; Handle_FireDelay();
}

bool ASandboxWeapon_Base::CheckAmmo()
{
    return _AmmoPerFire <= _CurrentAmmo;
}

void ASandboxWeapon_Base::PlaySound(USoundBase* sound)
{
    UGameplayStatics::PlaySoundAtLocation(this, sound,
        GetActorLocation());
}

void ASandboxWeapon_Base::UpdateAmmo()
{
    _CurrentAmmo -= _AmmoPerFire;
}
```

On the topic of being specific, we now have a `LinetraceOneShot` function in place of the `ProcessFiring` function from before. While the utility of this, specifying that it is one shot, will become apparent later, this function is one of many ways to handle the gameplay logic of the gun firing. It acts as a wrapper for the line trace function working off a direction vector the user must pass in. This then applies damage to whatever it has hit through the standard Unreal method:

We can now add that to the sandbox weapon:

SandboxWeapon_Base.cpp LinetraceOneShot function definition

```cpp
void ASandboxWeapon_Base::LinetraceOneShot(FVector direction)
{
    FHitResult hit(ForceInit);
```

```
    FVector start = _FirePoint->GetComponentLocation();
    FVector end = start + (direction * _Range);
    if(!UKismetSystemLibrary::LineTraceSingle(GetWorld(),
        start, end,
        UEngineTypes::ConvertToTraceType(ECC_Visibility),
        false,
        {this, GetOwner()},
        EDrawDebugTrace::ForDuration,
        hit,
        true,
        FLinearColor::Red, FLinearColor::Green, 5))
            return;

    UGameplayStatics::ApplyDamage(hit.GetActor(),
        _DamagePerHit, GetInstigatorController(),
            GetOwner(), UDamageType::StaticClass());
}
```

Now that our sandbox base is kitted out with a basic suite of tools; we can make the Blueprint child weapons equivalent to the template pattern by using the Sandbox_Fire function. Comparing the Blueprint implementations of each pattern should show a lot of the differences in approach and how each can be used for the same system with differing approaches.

Creating child classes

Just like with the template pattern, we are going to create two child classes of the Sandbox Weapon Base in a Blueprint. To do this, we follow the same steps except when searching in the **ALL CLASSES** list, we now type sandbox and select **SandboxWeapon_Base**. Name the two child classes BP_SandboxPistol and BP_SandboxShotgun.

With those created we can move on to setting them up. Unlike with the template pattern, we don't need to create the functional behaviors. Instead, we determine the order (and reuse) of the functions in the parent.

Sandbox pistol

Once again, we begin with the pistol as this is probably the simplest weapon type to implement. First, we do all of our checks to make sure the weapon can fire, as follows:

1. Override **Sandbox Fire Function** using the **Override** list.
2. Add a Branch based on the CanFire Boolean.
3. From the **True** pin, call Check Ammo and add a second Branch based on the **Return Value**

Figure 9.9: The start of the Sandbox Fire Event

Next, we'll add the logic to the **False** output from the **Branch** node, calling **Reload** and playing a sound (sounds haven't been implemented in the parent class as that is something that will typically be different per weapon):

1. Call the Reload function.

2. Add a Play Sound node, selecting the **Gizmo_Handle_Clicked** sound from the dropdown.

Figure 9.10: The false logic for the Sandbox Fire Event

Lastly, we can add the True logic, which essentially fires the weapon. As the functional behavior already exists in the parent, we simply need to call the functions and provide the required inputs. We're going to do this in two chunks. The first will play a sound before performing the line trace and updating the ammo. The second chunk will use a timer to call the function to manage the rate of fire:

1. Add a Play Sound node, selecting the VR_Teleport sound once again.

2. Call the Linetrace One Shot function, providing it with the Forward Vector of the **Fire Point** component.

3. Call Update Ammo.

Figure 9.11: The start of the true logic that fires the sandbox pistol

4. Add a Set Timer by Function Name node.

5. Type Handle_FireDelay in for **Function Name**.

6. Divide 1 by the **RoF** float variable and connect it to the **Time** pin of the **Set Timer by Function Name** node.

7. Promote the **Return Value** to a variable and call it TimerFireDelay.

Figure 9.12: The Hand_FireDelay timer added to the end of the chain

With the pistol complete, test it the same as we did with the two template weapons and then we can move on to implementing the shotgun using the Sandbox pattern.

Sandbox shotgun

The shotgun follows much of the same setup as the pistol except for where we once again need to complete multiple line traces. The parent has a single line trace that requires a forward vector input. So, just like in the template example, we will use a **For Loop** and once again utilize the **Random Unit Vector in Cone in Degrees** node when providing the forward vector input to the **Linetrace One Shot** function:

1. Start by copying the logic from the pistol. This just saves us a little bit of time.

2. Disconnect both sides of the **Linetrace One Shot** node by holding down *Alt* and clicking on the pins.

3. Add in a For Each loop just as we did in the template shotgun, connecting it after the **Play Sound** node in the True logic and once again using the Number of Pellets integer variable.

4. For the **Direction** vector input on the **Linetrace One Shot** node, set up **Random Unit Vector in Cone in Degrees** in the same way we did for the template shotgun, as shown in *Figure 9.13*

Figure 9.13: The For Loop's Loop Body logic for the BP_SandboxShotgun

Now we have the multiple line traces happening, we need to add back in the rest of the firing chain, that is, the Update Ammo function and Fire Delay Timer. Since these steps need to be done after we fire, they need to be connected to the **Completed** pin on the **For Loop** node, not part of the **Loop Body.**

Move the nodes from the previous steps around to make a clear path for the connection from the **Completed** pin and connect the other nodes back in, as shown in *Figure 9.14*.

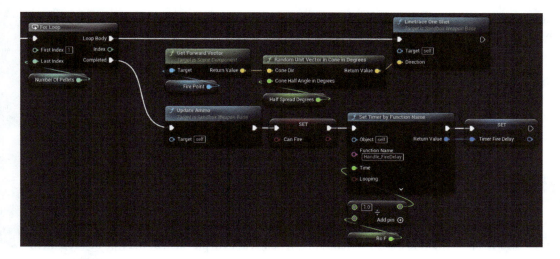

Figure 9.14: The Update Ammo and Timer connected to the Completed pin

As before, test this out by selecting the weapon on the Elite Unit in the level.

Provided everything works as it should, we can then move on to exploring the type object pattern, which allows us to easily expand the content available in a game.

Type object pattern

If you need a fast way of creating many variants of something in your game as a form of expanding the content available to players, then the type object pattern is for you. Type object takes the ideas of implicit and explicit data we covered as part of the flyweight pattern back in *Chapter 3* and expands it into the world of gameplay. The principle is the same: we separate out all data that is common across all instances of a type, but instead of just linking to it from everywhere, we mix it up and produce lots of variations of this data. The result is a connected web of objects that all have the same functionality but vary in which set of implicit data they use.

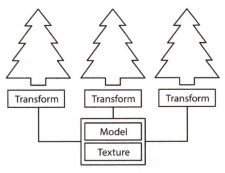

Figure 9.15: Diagram from *Chapter 3* where we discussed the Flyweight pattern

Figure 9.15 shows the Flyweight pattern saving space by storing implicit data about the idea of a tree in one place in memory.

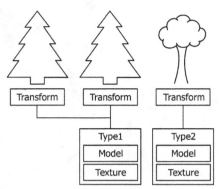

Figure 9.16: Diagram showing the expansion of the Flyweight pattern provided by the type object pattern

Expanding the pattern from Flyweight to type object in *Figure 9.16*, we define more types as different sets of the implicit data. This diagram makes it clear the cost of this added expandability will be memory. Now only objects sharing a type can be batched together, so the number of batch calls increases but we are still way beyond defining every tree explicitly.

This is not the only way of achieving this outcome though. We could keep our Flyweight pattern and make a new class for each of our types. That would work, so why don't we do that? Because code takes up space. In the next chapter, we'll cover the idea of data locality, it builds on the knowledge that instructions in code, such as values we manipulate, take up space in memory. Duplicating a class also duplicates the code that needs to be stored to run, not to mention the overhead cost of defining and storing a new class. Our aim is to make an elegant solution that streamlines the content creation process and increases efficiency.

To make the type object pattern work in Unreal, we need to store a collection of data as an asset that can be loaded into the RAM at runtime for referencing. This could be achieved with writing text files or some other structured file format like JSON or XML. The problem there is that designers need to have the data files open in a separate text editor to make changes, and the process of saving and reloading the editor preview can become tedious when many small changes are made iteratively. Thankfully, Unreal Engine gives us a few options of built-in structures we can use. We will be looking at Variants and the Variant Manager, Data Assets, and Data Tables. There are undoubtedly more ways to make this work but these three should cover most of the general implementations of the type object pattern.

Variants

The **Variant Manager** is a tool within Unreal Engine that allows us to create multiple swapable Variants of actors in a level. The Variants store values for Actor properties and can also call functions from within the actor when the Variant is selected.

The Variants are held within the level by a **Level Variant Sets Actor**, which links to a **Level Variant Sets** asset where the various actors, their properties to change, and the functions within them to call, are all stored.

Variants are activated either via the **Variant Manager Panel** while in the editor (as shown in *Figure 9.17*) or via Blueprint functions at runtime.

Figure 9.17 The Variant Manager panel

Variants and the Variant Manager are more commonly used in interactive experiences such as architectural visualization projects or product-based applications such as car configurators.

The Variant Manager approach could be used for a character creation screen like those seen in RPG-type games such as World of Warcraft or Cyberpunk 2077, or the modular character/vehicle setup systems in games such as Fortnite, Fall Guys, or Rocket League, where the player can select different parts to be added to their character.

The drawback with the Variant Manager approach for ever-expanding games such as Fortnite or Fall Guys is the time required to create variants and fully set them up. An in-game character customization system would be much more suited to using a data table approach, where additional data can quickly be added in a table editor or imported from a spreadsheet. Let's take a look at this next.

Data Tables

Data Tables are an asset type created for the purpose of storing values with no functionality. We define them in our project with a struct, each member becoming a column in the table. Each row of the table then represents a set of values we call a type to fit in with our pattern. Data Tables are designed for large-scale storage of data in one place to be queried when necessary. They can store values pointing to other assets, but these are cumbersome to work with in this form and are generally avoided in favor of storing everything as primitive types. This soft limit of primitive data means that, although possible, working with nested information is advised against. This could affect systems where one customization affects what is available in a level below.

The main benefit of this asset is its ability to import directly from a CSV or JSON file, making it the preferred method of balancing for designers with external tools. The flipside of this is because everything is stored together, even if you aren't using all the types in a given scene, all the data will still be loaded into memory, making it less viable for high-level types and more applicable to utility-type patterns such as quest systems that are universal and always need to be loaded. There is also an argument for using Data Tables for language localization, as with all the types in one place, searches for specific records become way easier and faster.

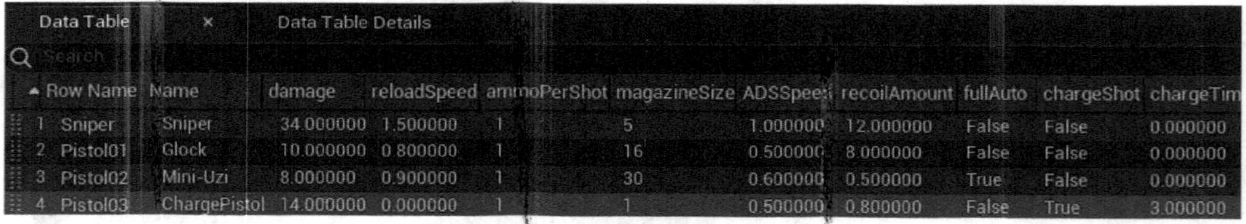

▲ Row Name	Name	damage	reloadSpeed	ammoPerShot	magazineSize	ADSSpeed	recoilAmount	fullAuto	chargeShot	chargeTim
1 Sniper	Sniper	34.000000	1.500000	1	5	1.000000	12.000000	False	False	0.000000
2 Pistol01	Glock	10.000000	0.800000	1	16	0.500000	8.000000	False	False	0.000000
3 Pistol02	Mini-Uzi	8.000000	0.900000	1	30	0.600000	0.500000	True	False	0.000000
4 Pistol03	ChargePistol	14.000000	0.000000	1	1	0.500000	0.800000	False	True	3.000000

Figure 9.18: Screenshot of an example Data Table with weapon balancing data

Figure 9.18 shows a Data Table in use for balancing the different enemy types in a game where the only changes are stat values as the enemy rank increases. This could have been imported from an external balancing tool as all the values are primitive types.

Data Assets

A Data Asset takes the idea of Data Tables and breaks it down into individual rows. Each row then becomes its own asset that can be created and managed via the editor. When we define a Data Asset it looks very much like a struct definition but on instancing it, we do not get a new item in the world. Instead, we get a new Data Asset instance in the project, similar to the way materials and material instances work. Once a Data Asset instance is referenced in active code, that instance is loaded into memory, just like a texture or static mech asset. Due to the editor being responsible for the creation and management of Data Assets, they have easy tools for holding references to other assets. This makes them useful for defining type data for high-level class specialization. They also deal with nested information well, as if it has been serialized properly, the editor has space to show the drop-down menus where sub-values can be manipulated.

The best way to explain how to use Data Assets for a type object pattern is to make something with them. So, let's turn our attention back to the project we have been working on through this chapter to implement some Data Assets:

1. Start by defining a new type of Data Asset by right-clicking in the C++ folder. Make sure to do this from the editor as there are no templates available within Rider, meaning you'll have to make a lot of unnecessary changes.

2. When selecting the base class to inherit from, choose **UDataAsset** from the **All Classes** menu.

3. Then give it a name. We have called ours EnemyType.

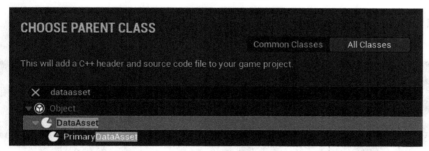

Figure 9.19: Class creation menu setup to make a new Data Asset

Next, we consider what data we need to store inside the asset. There could be any number of things we want to vary between different enemy types. The key is to make sure you only store data specifically about the enemy here, and not the other classes that could be different within the enemy class, such as the weapon fire rate. Weapons would need another parallel implementation of the type object pattern, which could even be nested within this one with the weapon type defined within the enemy type:

1. Give the class the BlueprintType property in the UCLASS block.

2. We are only changing the health and materials on the enemy with its type, so add a float for health and a couple of object pointers to the UmaterialInstance variables.

3. Make sure each variable has the EditAnywhere and BlueprintReadWrite property specifiers, and we are all done with the C++ side:

EnemyType.h exerpt

```cpp
UCLASS(BlueprintType)
class RTS_AI_API UEnemyType : public UDataAsset
{
    GENERATED_BODY()


public:

    UPROPERTY(EditAnywhere, BlueprintReadWrite)
```

```
        float _Health;

        UPROPERTY(EditAnywhere, BlueprintReadWrite)
        TObjectPtr<UMaterialInstance> _Material1;

        UPROPERTY(EditAnywhere, BlueprintReadWrite)
        TObjectPtr<UMaterialInstance> _Material2;
};
```

The next step is to define some Data Assets from the template we have just created:

1. Build back into the editor and right-click on the **RTS/Data** folder.

2. Select **Miscellaneous > Data Asset**.

Figure 9.20: The Data Asset option is in the right-click menu

3. Select **EnemyType** from the list of classes to be the parent.

4. Give the new asset a name that represents the type we are making (we have gone with Enemy_ Basic) and then open it up.

5. Set the health value to 100 and select **MI_EnemyUnit_01** and **MI_EnemyUnit_02** for the material slots.

The last step is to apply these values to the unit on spawn:

1. Open **BP_EnemyUnit**, found in the **RTS/Blueprints** folder.

2. Add a new variable of type **EnemyType** and make it editable by clicking the eye icon, as shown in *Figure 9.21*. Then set its default value to **Enemy_Basic**.

Figure 9.21: The BP_EnemyUnit variables

3. Add a **BeginPlay** event node to the event graph.

4. Check whether the new **TypeData** variable has a value.

5. Then place each value from the **TypeData** variable into its relevant place as shown in *Figure 9.22*.

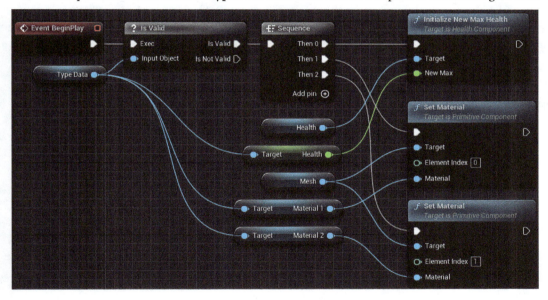

Figure 9.22 Screenshot showing BP_EnemyUnit event graph applying Type Data

Doing all this should make nothing appear or behave differently, but it does now give us the flexibility to define new types of enemy with different materials in the Content Browser. We can then set spawned enemies to be different types using the editable variable. As an extension to this, see if you can define a new enemy type with 200 health that uses the B versions of the material instances we used for the basic type. These can be found in the same folder as the materials we were using before but are tinted blue. Another extension you could try is to set the type variable to expose on spawn and dynamically spawn a few enemies in the level. This will allow you to set the type in the code, hopefully demonstrating the power of this system.

We've explored the ways in which the type object can be utilized in gameplay to facilitate large numbers of variants using three different approaches in Unreal Engine: Variants, Data Tables, and Data Assets. With the completion of our type object example, we have come to the end of our exploration of patterns that can be used to build structured solutions.

Summary

The three patterns we have covered in this chapter are the most widely used structural patterns. Use of the first two is a personal preference, as we saw they can both be used to achieve the same thing in different ways. Template and subclass sandbox are also being superseded by other techniques in modern code, such as interfaces and modular design, but understanding where they came from and the workflow they encourage is useful. Template and subclass Sandbox both highlight the need to constrain designers with limited access to ensure the maintainability of the codebase. The type object pattern, on the other hand, is one of the most useful patterns in game development with widespread application across all aspects of game design. Its utility in allowing artists, designers, and programmers to work together is invaluable.

In the next chapter, we will dive into a few patterns that we can apply once we have a working game to improve our performance using the concepts of object pooling, data locality, and dirty flags.

10

Optimization through Patterns

In this last chapter, we are going to discuss the last thing we should think about before releasing our games: optimization. Optimization patterns are designed to leave our code functioning as it was before but in a faster, more elegant way that impacts our hardware less. This chapter is quite wordy, but the underlying principles that guide these patterns require a certain understanding of how the hardware resources at our disposal work. By the end, we will have covered everything from how to help the CPU do its job better to making a system you can plug into any game to make it potentially faster at runtime.

The patterns making this possible are the following:

- **Dirty Flag**, which focuses on reducing the number of times we need to update calculated values.

- **Data Locality**, which concerns optimizing the code layout to work with the way the CPU's memory works. As a description, this sounds much more complicated than the reality of the application.

- **Object Pooling**, where we offset as much of the heavy memory allocation processing to the start of the game, where it can be excused under a loading screen, so as not to impact runtime efficiency.

So, in this chapter, we will cover the following topics:

- Using dirty flags to reduce unnecessary processing

- How data locality affects code efficiency

- Object pooling our resources to save time later

Technical requirements

The starting point for this chapter can really be from any project, but we have a branch of the GitHub repository that carries on from where *Chapter 9* left off. This provides a set of systems we can integrate with the Object Pooler we will be building. You can find this starting point in the *Chapter 10* branch here:

```
https://github.com/PacktPublishing/Game-Development-Patterns-with-
Unreal-Engine-5/tree/main/Chapter10
```

Using dirty flags to reduce unnecessary processing

Dirty flag involves updating values only when they are needed. The best explanation of how it works comes in the context of base-level engine development and the transform hierarchy. When you set a local location on a transform, you are indicating that you want to be x, y, and z units away from the parent's location. This is easy enough to update, but in doing this, we are also changing the transform's world space location. It is easy to calculate the matrix that will deal with this local-to-world space conversion, then multiply our vector by it; that process doesn't even cost many resources. Then, we must remember that this is a hierarchy. What if we were moving the root of a tree that is hundreds of transforms deep? Not a great position to be in for multiple reasons, but if the parent of a transform moves, then the child moves with it, and so on, recursively. This presents a lot of world space value transforms that need to be updated every time any parent in the tree is changed.

To make the utility of dirty flag easier to quantify, we can consider a hypothetical situation where we have a hierarchy of transforms that is 100 parent-child connections deep. We want to move each of them like they are a chain with a torque ripple. Starting at the top of the hierarchy and moving down, at every step, we update the position of the transform at that level to a new local location defined by some periodic function. With each local update, we also update the local-to-world matrix for every transform lower in the hierarchy as they will have moved in world space, as shown in *Figure 10.1*. This would require (101-n) matrix updates at each step, which means to move the entire hierarchy of 100 transforms, we will end up with the 100th triangular number, which is 5,050. I think we can safely say that's ridiculous and there has to be a better way. Consider the utility of the work done. Why are we updating these transforms? So that something else can read its world space location and get an accurate, up-to-date value. Have we read the world space location at any point in this algorithm? No. The function for setting the local location doesn't need the world space location. So, do we need to update the local to world matrices of these objects? Not until something else needs us to or the end of the frame is reached. For the best case, that means we could get away with only 100 matrix updates at the very end. That is the purpose of the dirty flag pattern.

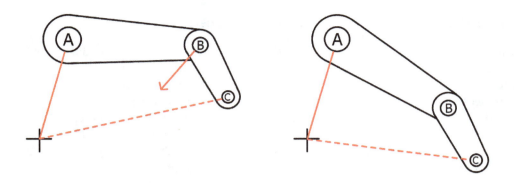

Figure 10.1 – Diagram showing a local location change affecting the world space location of its child

Now, let's look at the application of dirty flags.

Application of dirty flags

In practice, this whole pattern is just a Boolean value; when the object is considered *dirty*, the Boolean is one value, and when it is *clean*, it is the other. Which way around that is doesn't matter, as long as it's consistent with the naming. An object is dirty if it has pending changes that still need to be made to the values it represents. In our example, we postpone the local-to-world matrix update until something requests the matrix or the end of the frame is reached. During the time between the change to the local location and the matrix update, the transform is considered dirty.

The mechanics of how this is dealt with then extend beyond the Boolean value. On raising a dirty flag, that object is added to a relevant static dirty list of objects. For our transform, that means it, and all the transforms below it in the hierarchy, will be added recursively to a static list of other dirty transforms. We then have some cleaning function that describes how to go from a dirty state to a clean state. With the transform, that would be a function that calculates the updated local-to-world matrix.

That example is only something you'd have to be worried about if you were designing your own engine, but the dirty flag pattern can be applied to great effect anywhere you have a value that may be updated several times before it is needed. You can find this pattern in **Unreal Motion Graphics (UMG)** layouts, which have a hierarchical positioning dependency just like transforms, and the `destroy` command, which marks actors for destruction at the end of the frame due to the trickle-down effects of removing them. There will be cases where you are storing data that needs to be displayed to the UI and the dirty flag pattern can reduce the number of times you tell the UI to update per frame; it should only happen once at most per frame.

Next, we'll go deeper, from a one-variable pattern to a guiding principle of optimization that we can apply anywhere.

How data locality affects code efficiency

This is a simple concept that requires very little to implement. We all actively think about how variables take up memory. We tend to forget that instructions also take up memory. They must be loaded from storage into faster memory, then into the CPU to be executed. CPUs try to make things run quicker by leveraging the fact that they have very fast, very small storage within them called the cache. The cache allows the CPU to pre-load instructions ready for execution and store its state in case of temporary context switching. This pre-loading behavior is necessary for the CPU to run at its most efficient as while there has been a technology race in CPU speeds, that hasn't been mirrored in the world of RAM. You might be able to store massive programs entirely in your RAM but the bus speed of the motherboard limits how many instructions can be sent to the CPU per second. When we reach the bottleneck, it doesn't matter how fast the CPU cores are at calculating results, as they would spend most of their time idle, waiting for instruction. Pre-loading provides a mechanism that has the potential to fix this by sending large chunks of instructions to the CPU cache for processing. We say *potential to fix* as when pulling instructions from RAM, there is no way of knowing which instructions will be next. That information is in the instruction daisy chain and can only be accessed once the work is done. This means that the contents of the CPU cache are entirely dictated by the geography of our system architecture.

That's lots of technical terms, so let's explain it with an analogy. Imagine working in a factory where your job is to build flatpack furniture. You're really fast at your job when you have the materials and can put together each piece lighting fast. The catch is, you can only see one instruction at a time, and when you need pieces, you must request them from a porter. These resources, such as panels and screws, are stored in a warehouse miles away. When you request something, the porter spends a day traveling to and from the warehouse, which means you can only get one instruction completed per day regardless of how fast you work. Most of your time is spent staring at the wall. In this example, you are the CPU, executing instructions, and the porter is the data bus, ferrying instructions from RAM to be executed.

One day, a new manager is brought in who decides to revamp the process. They change the material request process so that now, when the porter gets the required item, they also get everything else within arm's reach. This bundle of screws and panels is then dumped on the ground in your workstation. In real terms, we call this the CPU cache, a tiny amount of extremely fast memory within the CPU. The benefit of this is that when you get your next instruction, there is a chance you already have the required materials next to you. If not, then all the materials need to be taken back and a new bunch collected. It then stands to reason that if the warehouse is organized so that materials that are often requested together are placed near each other, the porter is more likely to take the correct materials for the next instruction as well. Nothing needs to change about the porter's knowledge of the situation or skills, simply proper planning at the start to achieve an efficient outcome.

As programmers, we can be the warehouse manager in that example, making sure that the data a function requires is physically close to that function so that when the CPU request for resources comes in, the cache is more likely to fill with useful data. When the CPU can execute from the cache, that is called a cache hit. Likewise, when the data needs to be requested, that is a cache miss. We want to achieve as many cache hits as possible to reduce the number of times the cache needs to be refilled. The gains from achieving high levels of cache hits are surprising; sometimes, they can be up to 50 times faster due to organizing data effectively.

We're going to have a look at two methods for implementing data locality as a principle but there are doubtless others that, now that you understand the problem, will make more sense as implementations. Let's look at the two methods.

Hot/cold splitting

The first technique is very similar to how the type object pattern from *Chapter 9* needed to consider implicit and explicit data, but instead of thinking about what the values are defining, we look at how frequently they are accessed. The go-to example of this would be an NPC in a game with loot drops. The NPC's health is accessed regularly as they heal and take damage over their life cycle, whereas the loot table, which describes what items they drop on, is accessed once at the end of the object's life. We can classify the frequently accessed data as hot; this can stay in the object as member variables. The more single-use data, such as loot tables, is then marked as cold and separated off into a struct, held inside the object as a pointer.

Why do all this? It has to do with the size of the object when being pulled into the cache. When the object is pulled in, all data it directly contains makes up the amount of space it takes up in the cache. That means that all pointers effectively only take up the space of `uint64_t`. The data they point to is not necessarily loaded until it is directly accessed, as it is declared physically elsewhere, hence the pointer. Without separating our hot and cold data, as we described previously, our class takes up more cache memory than is necessary with data that is unlikely to be needed, increasing the chance of a cache miss.

Contiguous arrays

The second technique is using contiguous arrays of data. We know that there are two types of memory: stack and heap. Data locality is the main reason why stack memory is considered faster. Everything in the stack has been defined before the program runs and so it is neatly organized. Arrays of data are held in the stack and are defined together in one continuous line. This makes the CPU cache more efficient when looping over these elements as they have been stored physically closer to one another. This is part of the reason why data-oriented ECS is faster, as discussed back in *Chapter 4*. However, dynamic collections and pointers are declared in heap memory at runtime. We sacrifice that benefit of efficiency for the flexibility of defining data at a later point. Data on the heap uses whatever free space is available and because of this may end up defining multiple objects large distances from each other. *Figure 10.2* shows visually how storing an array of values instead of an array of pointers can make a difference to what is loaded into the cache. This is a concept to keep in mind when we implement an object pool later in this chapter. When we spawn objects, they are held as `TObjectPtrs` in a `TArray`. Could this array be made into a standard C++ array? What dynamic property would we have to sacrifice to do this? It would likely be dynamic sizing, but do you need that in your context?

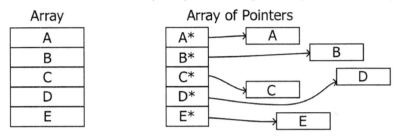

Figure 10.2 – Possible layout of an array of pointers versus an array of values in memory

So, is the solution to just use arrays of data all the time? Well, no. There are many situations where pointers and dynamic collections are still necessary within object-oriented programming. This is more a point to consider their usage, and if you can replace a dynamic collection with a static array, then do so. This point is especially important to remember in our last pattern of the book coming up next: Object Pooling.

Object pooling our resources to save time later

The last pattern we are discussing in this book is object pooling. This pattern aims to tackle one of the core problems with the CPU: allocating and deallocating memory is a slow process. Every time you spawn a new actor, the space it needs in memory must be reserved in the right subdivisions for each variable and handed back to the game process ready to receive data. Every time you delete an actor, that memory must be freed from all references and returned to the heap. For something such as a minigun spawning 3,000 projectiles per minute, that means a lot of allocation of the chunks for memory, which are all the same size. Object pooling is the practice of predicting this massive cost and offsetting it to a place where the lag it causes is not so noticeable. This for us means spawning all the projectiles we could possibly need at the start of the game and hiding them. When one is required, it is taken from the shelf of deactivated projectiles, teleported to the right position, and activated. Then, to preserve the pool's integrity, when it would have been destroyed, it is simply deactivated and returned to the shelf with the other pooled items. Although this pattern does make the one frame when the level has just loaded much worse, as we are spawning all the projectiles at once, we can disguise this under a loading screen. This offset can dramatically increase the processing speed during hectic gameplay sequences when there would have been a lot of spawning and destroying occurring. Our pool then stays active until the end of the level when all objects are destroyed together, again under a loading screen.

With the theory covered, let's make some object pools.

Implementing object pooling

There are a few ways we could look at implementing this and, realistically, if it works to spawn the objects you need in a place where you can access them, then it is the best method for you. The implementation options are as a world subsystem, level actor component, game mode component, or floating actor in the level.

World subsystem

Subsystems are Unreal's effort at implementing a standardized form of the singleton pattern we covered in *Chapter 8* with limited scope. This form it takes means that we can make an almost static class that we know will exist for the lifetime of whatever it is attached to. Subsystems are, however, not well protected with regard to access as anything with reference to their attached object can call functions on them. This is why they tend to be used for hidden logic behavior systems that run regardless of interaction. This results in most of the public functions on them being getters to get the state of something they are processing. There are five levels of subsystem that exist. Let's describe them in order of decreasing lifetime:

- **Engine**: Exists in both the editor and the game for the length of the executable running.

- **Editor**: Runs as an editor tool and will not build with the game.

- **Game instance**: Attached to `UGameInstance` and so exists for the play session while the executable is running. Only one instance can exist at a time.

- **Local player**: Matches the lifetime of the `ULocalPlayer` it is attached to and moves between levels in the same way. There is one instance per local player.

- **World**: Matches the lifetime of the `UWorld` it is attached to. There is one instance per `UWorld` that is currently loaded.

It is important to consider the scope of the system being created and match it to the parent that best describes its lifetime. For an object pooler, this would be a world subsystem as any objects spawned in a pool will exist within a world so when that world is unloaded, they will be too. If the system was made as a local player subsystem, this would break references when changing maps and possibly spawn items in menu worlds where they are unnecessary.

LevelScriptActor child

The `ALevelScriptActor` is what people know on the blueprint side as the level blueprint. It provides a place for level-specific code to execute. This can be useful for tutorials where the mechanics are built badly in order to introduce them slowly, or for map-based mechanics, such as the "Levolutions" in Battlefield 4, where each map has the ability to completely change if different conditions are met. What isn't advertised very well in the Unreal documentation is that we can change the level blueprint in the C++ layer. Simply create a new C++ child of `ALevelScriptActor` and add your code here. This new child can be where we set up systems for object pooling as the `ALevelScriptActor` exists in a hidden state for as long as our world exists and has easy access to anything else within the world outliner for that particular map. The downside of this is that every new map created in the editor comes with a level blueprint that already inherits from the base `ALevelScriptActor` class. This means every new map would have to have its level blueprint manually reparented to your custom C++ type, which could lead to a lot of admin with easily missable steps.

Game mode component

The game mode is a class that is guaranteed to be in every level, so there is the option to make the object pooler an actor component that is attached to it, or consolidate the behavior into the game mode inheritance hierarchy with a custom pooling game mode, which innately has the logic for running an object pool built in. This approach would require some diligence on the part of the designers as making a new level or prototyping a new game mode would require the component to be added or the correct parent to be selected; but it would make the implementation easy seeing as it is collected in one place and self-contained within an easily accessible system.

Floating actor

The last method of getting an object pooler to work is the simplest but least elegant solution: making it into an actor that you spawn into the level. The benefit of this is you can easily have multiple object pools for different things or segregate your object pools by area if you are dealing using world partition, the system we discussed back in *Chapter 3*. The setup is also simple as all the GUI for setting it up is collected into the details panel for that object pool. The reason we call this method inelegant is down to how it must be managed. With there being no central method for referencing it or making sure the required functions have been called, it leaves a lot up to the end user and therefore is error prone.

Making an object pool

Before we start, the object pooling pattern is probably the most useful pattern to have in a plugin that we can take between projects. So, anything we make here should probably be done as part of a new plugin, which we can make within rider using the right-click menu on the game project. Simply select **New Unreal Plugin...** from the **Add** menu, as shown in *Figure 10.3*, and call the plugin something sensible, such as `ObjectPooler`. Then, just be sure to add new classes for the object pooler under the new folder this creates in the **Source** directory.

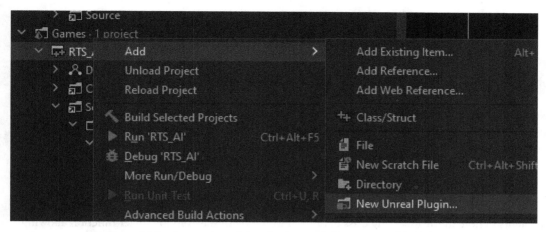

Figure 10.3 – Screenshot of the plugin creation process within Rider

Our first step is to make the struct that will define the attributes of a single type of pool. The code for this is presented next, but let's explain some of the key points. First, the `BlueprintType` property in the `USTRUCT` block, in combination with the `EditAnywhere` property specifiers, will allow the end user to change the pool behavior in the editor. There is also the constructor, which must give every property a value as a struct cannot be a `nullptr` in memory. Saving the `_ActorName` variable as an `FString` is done to make debugging easier, but if you prefer to save it as an `FName`, that works and will save some processing when the pool is warming up:

PooledObjectData.h struct

```cpp
USTRUCT(BlueprintType)
struct FPooledObjectData
{
    GENERATED_BODY()

    UPROPERTY(EditAnywhere)
    TSubclassOf<AActor> _ActorTemplate;
    UPROPERTY(EditAnywhere)
    int _PoolSize;
    UPROPERTY(EditAnywhere)
    bool _CanGrow;
    UPROPERTY(EditAnywhere)
    FString _ActorName;

    FPooledObjectData()
    {
        _ActorTemplate = nullptr;
```

```
        _PoolSize = 1;
        _CanGrow = false;
        _ActorName = "default";
    }
};
```

Next, we'll turn our attention to the component that will be on every object that came from the pool. We use an actor component instead of making a new child of AActor that must then be inherited from as it provides a clean separation between the object existing and doing what it needs to and the hook that attaches it back to the pool it came from. With this setup, we can dynamically spawn the component at runtime and attach it to the object when it joins the pool, keeping reference only to the component. This should make the pool totally class agnostic, improving its versatility.

Elements to note in the following class definition would be the custom initializer function allowing us to set the component up properly (more on that when we get to the object pool side) and the BlueprintCallable function used to recycle the actor. The recycle function is to be used instead of the standard Destroy on the actor as it will return its owning actor to the pool it came from. A useful extension you might want to add here would be to save the index of the pool it is supposed to return to. This will save some string comparisons later:

PooledObject.h

```
class AObjectPool;

UCLASS(ClassGroup=(Utility), meta=(BlueprintSpawnableComponent))
class RTS_AI_API UPooledObject : public UActorComponent
{
    GENERATED_BODY()
public:
    UPROPERTY(VisibleInstanceOnly, BlueprintReadOnly)
    bool _IsActive;

    void Init(AObjectPool* owner);
    UFUNCTION(BlueprintCallable)
    void RecycleSelf();
private:
    TObjectPtr<AObjectPool> _ObjectPool;

    virtual void OnComponentDestroyed(bool bDestroyingHierarchy)
    override;
};
```

The implementation of these functions is then very simple as most of the logic will be run in the pool itself. The only interesting point of note here is the `OnComponentDestroyed` override. This function removes the `RecycleSelf` function as a listener to a delegate on the pooler as a safety in case the pooler functionality is ignored, and the object is deleted in error:

PooledObject.cpp

```cpp
void UPooledObject::Init(AObjectPool* owner)
{
    _IsActive = false;
    _ObjectPool = owner;
}

void UPooledObject::RecycleSelf()
{
    _ObjectPool->RecyclePooledObject(this);
}

void UPooledObject::OnComponentDestroyed(
    bool bDestroyingHierarchy)
{
    _ObjectPool->OnPoolerCleanup.RemoveDynamic(this,
        &UPooledObject::RecycleSelf);
    Super::OnComponentDestroyed(bDestroyingHierarchy);
}
```

Now for the main event, the object pool itself. Breaking down the definition, we start with a new delegate type with no arguments. This exists as a tether for each of the objects taken from the pool. If we need to recall them due to a level change, we can broadcast this delegate to recycle all active objects. We then have the definition of a new struct type. This only exists as a workaround for the fact that the template collections inside Unreal do not cater to multi-dimensional arrays. We would like to store an array of pools that in themselves are arrays. So, to get around this limitation, we define a new struct type that will hold all the objects we consider to be a part of one pool:

ObjectPool.h excerpt part 1

```cpp
DECLARE_DYNAMIC_MULTICAST_DELEGATE(FPoolerCleanupSignature);

class UPooledObject;

USTRUCT(BlueprintType)
struct FSingleObjectPool
{
```

```
    GENERATED_BODY()

    UPROPERTY(VisibleInstanceOnly, BlueprintReadOnly)
    TArray<TObjectPtr<UPooledObject>> _PooledObjects;
};
```

Next is the object pooler class. An exception to the rule, we don't mark this as abstract. The reason for this is this actor just needs to exist. There is no need for any visual elements and so it can exist entirely on the C++ side, calling back to our separation rules for establishing the fuzzy layer in *Chapter 1*. The API includes functions for broadcasting the cleanup delegate, getting an object from the pool, and two methods for returning an object to the pool with either a `UPooledObject` component reference or a straight `AActor` reference. We'll go over why there are two later in the definitions. In the protected section, we need a `BeginPlay` override, an array of the data about the pools marked as `EditAnywhere` for designers to use the tool, and an array of the struct we made earlier to store a reference to every object this pool spawns. You could make this simpler by having a different object pool per object type, but that creates more actors than is necessary in the scene. Lastly, there is a private function for regenerating objects that have been deleted, leaving holes in the pool:

ObjectPool.h excerpt part 2

```cpp
UCLASS()
class RTS_AI_API AObjectPool : public AActor
{
    GENERATED_BODY()
public:
    UPROPERTY()
    FPoolerCleanupSignature OnPoolerCleanup;

    UFUNCTION(BlueprintCallable)
    void Broadcast_PoolerCleanup();

    UFUNCTION(BlueprintCallable)
    AActor* GetPooledActor(FString name);

    UFUNCTION(BlueprintCallable)
    void RecyclePooledObject(UPooledObject* poolCompRef);

    UFUNCTION(BlueprintCallable)
    void RecycleActor(AActor* pooledActor);

protected:
    virtual void BeginPlay() override;

    UPROPERTY(EditAnywhere, BlueprintReadWrite)
```

```
    TArray<FPooledObjectData> _PooledObjectData;

    UPROPERTY(VisibleInstanceOnly, BlueprintReadWrite)
    TArray<FSingleObjectPool> _Pools;

private:
    void RegenItem(int poolIndex, int positionIndex);
};
```

With everything declared, we can move on to the definitions of our functions. To start, we have the broadcast function, which works as its name suggests, and the `BeginPlay` override for *warming up* the pool by spawning all the requested objects. Each pool iterates over the predefined number of times spawning new actors in the world. The code here names them and crucially adds an instance of the `UPooledObject` component to them. Having the pooler add this component dynamically means that the person who developed the actor being pooled didn't need to know this was going to be added as a pooled class. This implementation uses `NewObject<>`, `RegisterComponent`, and `AddInstanceComponent` to create and add the component to the new actor as we are in runtime, and we would like to see the component in the actor details panel for debugging purposes. The new component needs its initialization function running before we hide it from view, disable its collision, and stop it from executing:

ObjectPool.cpp excerpt part 1

```cpp
void AObjectPool::Broadcast_PoolerCleanup() {
    OnPoolerCleanup.Broadcast();
}

void AObjectPool::BeginPlay() {
    Super::BeginPlay();
    FActorSpawnParameters spawnParams;
    for(int poolIndex = 0; poolIndex <
        _PooledObjectData.Num(); poolIndex++)
    {
        FSingleObjectPool currentPool;
        spawnParams.Name =
            FName(_PooledObjectData[poolIndex]._ActorName);
        spawnParams.NameMode =
            FActorSpawnParameters::ESpawnActorNameMode:: Requested;
        spawnParams.SpawnCollisionHandlingOverride =
            ESpawnActorCollisionHandlingMethod::AlwaysSpawn;
        for(int objectIndex = 0; objectIndex <
            _PooledObjectData[poolIndex]._PoolSize;
                objectIndex++)
```

```
        {

                AActor* spawnedActor = GetWorld()->
                    SpawnActor(_PooledObjectData[poolIndex].
                        _ActorTemplate, &FVector::ZeroVector,
                            &FRotator::ZeroRotator,
                                spawnParams);
                UPooledObject* poolComp =
                    NewObject<UPooledObject>(spawnedActor);
                poolComp->RegisterComponent();
                spawnedActor->AddInstanceComponent(poolComp);
                poolComp->Init(this);
                currentPool._PooledObjects.Add(poolComp);
                spawnedActor->SetActorHiddenInGame(true);
                spawnedActor->SetActorEnableCollision(false);
                spawnedActor->SetActorTickEnabled(false);
                spawnedActor->AttachToActor(this,
                    FAttachmentTransformRules::
                        SnapToTargetNotIncludingScale);
            }
        _Pools.Add(currentPool);
    }
}
```

The method for getting an object from the pool has been made with an FString argument to make it as foolproof as possible, but it is advised that you establish an enum type that can be used to reference the pools as indexes. In its current form, it goes through a few steps:

1. Finds the index of the pool that matches the input string, returning out if one isn't found.

2. Loops through the objects in the found pool to find the next object, which is marked as inactive:

    I.   If a nullptr is found, then regenerate an object at that position and return it as it will be available.

    II. If the end of the list is reached, then check whether the pool is allowed to grow. If it can, then make and return the new item; otherwise, it would be sensible to output a warning so that the designers know the pool probably needs expanding.

In the following code, the section for returning the object, if it is new or existing, is repeated due to the slightly different situations where a new object must have the component added and initialized but then does not need to be deactivated:

ObjectPool.cpp excerpt part 2

```cpp
AActor* AObjectPool::GetPooledActor(FString name)
{
    int poolCount = _Pools.Num();
    int currentPool = -1;
    for(int i = 0; i < poolCount; i++)
    {
        if(_PooledObjectData[i]._ActorName == name)
        {
            currentPool = i;
            break;
        }
    }
    if(currentPool == -1) { return nullptr; }

    int pooledObjectCount =
        _Pools[currentPool]._PooledObjects.Num();
    int firstAvailable = -1;
    for(int i = 0; i < pooledObjectCount; i++)
    {
        if(_Pools[currentPool]._PooledObjects[i] !=
            nullptr)
        {
            if(!_Pools[currentPool]._PooledObjects[i]->
                _IsActive)
            {
                firstAvailable = i;
                break;
            }
        }
        else
        {
            RegenItem(currentPool, i);
            firstAvailable = i;
            break;
        }
    }

    if(firstAvailable >= 0)
```

```
    {
        UPooledObject* toReturn =
            _Pools[currentPool]._PooledObjects[firstAvailable];
        toReturn->_IsActive = true;
        OnPoolerCleanup.AddUniqueDynamic(toReturn,
            &UPooledObject::RecycleSelf);
        AActor* toReturnActor = toReturn->GetOwner();
        toReturnActor->SetActorHiddenInGame(false);
        toReturnActor->SetActorEnableCollision(true);
        toReturnActor->SetActorTickEnabled(true);
        toReturnActor->AttachToActor(nullptr,
            FAttachmentTransformRules::
                SnapToTargetNotIncludingScale);
        return toReturnActor;
    }

    if(!_PooledObjectData[currentPool]._CanGrow) { return
        nullptr; }

    FActorSpawnParameters spawnParams;
    spawnParams.Name =
        FName(_PooledObjectData[currentPool]._ActorName);
    spawnParams.NameMode =
        FActorSpawnParameters::ESpawnActorNameMode::
            Requested;
    spawnParams.SpawnCollisionHandlingOverride =
        ESpawnActorCollisionHandlingMethod::AlwaysSpawn;
    AActor* spawnedActor = GetWorld()->
        SpawnActor(_PooledObjectData[currentPool].
            _ActorTemplate, &FVector::ZeroVector,
                &FRotator::ZeroRotator, spawnParams);
    UPooledObject* poolComp =
        NewObject<UPooledObject>(spawnedActor);
    poolComp->RegisterComponent();
    spawnedActor->AddInstanceComponent(poolComp);
    poolComp->Init(this);
    _Pools[currentPool]._PooledObjects.Add(poolComp);
    poolComp->_IsActive = true;
    OnPoolerCleanup.AddUniqueDynamic(poolComp,
        &UPooledObject::RecycleSelf);
    return spawnedActor;
}
```

The two recycling functions act as a way to do overloading with UFUNCTIONs. Unreal does not support this standard C++ practice out of the box, and so we must define new functions for each as a workaround. In this case, the RecycleActor function tries to get a UPooledObject component reference from the input actor. It may be worth adding a summary comment above this function, with triple forward slashes, letting the user know that it may fail and a better method would be to use the UPooledObject version. If it succeeds, it then calls the RecyclePooledObject function with this new information. Otherwise, it currently does nothing, but this may be a good place to log out the situation as a warning and maybe have the function return a Boolean value on successful recycling as feedback on the action. The main recycling function simply returns the object to its initial disabled and hidden state in the pool, resetting the _IsActive flag in the pooled component:

ObjectPool.cpp excerpt part 3

```
void AObjectPool::RecyclePooledObject(UPooledObject* poolCompRef)
{
    OnPoolerCleanup.RemoveDynamic(poolCompRef,
        &UPooledObject::RecycleSelf);
    poolCompRef->_IsActive = false;
    AActor* returningActor = poolCompRef->GetOwner();
    returningActor->SetActorHiddenInGame(true);
    returningActor->SetActorEnableCollision(false);
    returningActor->SetActorTickEnabled(false);
    returningActor->AttachToActor(this,
        FAttachmentTransformRules::SnapToTargetNotIncludingScale);
}

void AObjectPool::RecycleActor(AActor* pooledActor)
{
    if(UPooledObject* poolCompRef =
        Cast<UPooledObject>(pooledActor->
            GetComponentByClass(UPooledObject::StaticClass())))
    {
        RecyclePooledObject(poolCompRef);
    }
}
```

The last function rounding out our object pooler is a function for regenerating items. This could maybe be separated better to make it more useful, in the `GetPooledActor` function, but as it stands, this follows the standard object generation as in the `BeginPlay` method, just with a twist. It uses indexes to add an object to a specific place in the pooled array. There is a lot of room for improvement with this function to make it more versatile, but that is left to your implementation's needs:

ObjectPool.cpp excerpt part 4

```cpp
void AObjectPool::RegenItem(int poolIndex, int positionIndex)
{
    FActorSpawnParameters spawnParams;
    spawnParams.Name =
        FName(_PooledObjectData[poolIndex]._ActorName);
    spawnParams.NameMode =
        FActorSpawnParameters::ESpawnActorNameMode::Requested;
    spawnParams.SpawnCollisionHandlingOverride =
        ESpawnActorCollisionHandlingMethod::AlwaysSpawn;
    AActor* spawnedActor = GetWorld()->
        SpawnActor(_PooledObjectData[poolIndex].
            _ActorTemplate, &FVector::ZeroVector,
                &FRotator::ZeroRotator, spawnParams);
    UPooledObject* poolComp =
        NewObject<UPooledObject>(spawnedActor);
    poolComp->RegisterComponent();
    spawnedActor->AddInstanceComponent(poolComp);
    poolComp->Init(this);
    _Pools[poolIndex]._PooledObjects.Insert(poolComp,
        positionIndex);
    spawnedActor->SetActorHiddenInGame(true);
    spawnedActor->SetActorEnableCollision(false);
    spawnedActor->SetActorTickEnabled(false);
    spawnedActor->AttachToActor(this, FAttachmentTransformRules::
    SnapToTargetNotIncludingScale);
}
```

As stated a few times previously, this object pooler will do the job, but it is very basic in its utility. There are many extensions that you could, and probably should, consider, such as having pool groups so that objects are pooled based on the requested groups from the level or making it into a world subsystem that is universal to that world, allowing easy setup via the **Project Settings** panel. However, the principle use of it stays the same: to offset the cost of spawning to the start of a level, where it can be hidden under a loading screen.

Using what we have created in its current form is quite simple. Simply drag an instance of the object pooler into your world from the Project panel and set up its data variable in the details panel. Once the game starts, it will spawn all the required objects in and hide them. To get an object, all you need to do is obtain a reference to the pooler somehow and call the GetPooledObject function, as shown in *Figure 10.4*.

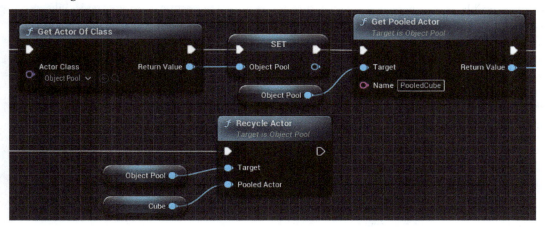

Figure 10.4 – Screenshot of the blueprint usage of the object pooler

With that, we are at the end of our journey through object pooling. If you have been following along, you will not only have an object pooler that you can migrate between projects but also an idea of how you can remake and improve it to suit specific needs as and when required. Not only this, but the end of this section also brings us to the end of the chapter and the book. Even though this wasn't specifically designed as a book to be read in order, from cover to cover, if you have been following this journey from the beginning, then you have a good set of practical skills and templates for how to improve your code in numerous ways. There will be some more parting words of wisdom after this, but let's round this chapter content out by saying that it is more important to get something working than to build it exactly correct from the start, which is why the term refactoring exists.

Summary

In this last chapter, we have covered three patterns that will boost the efficiency of your game, if implemented correctly. In game development, optimization should not be something you consider until it becomes a problem. It is far more important that you get something working first. Data locality should probably be considered as a first measure as it requires the least refactoring of code. Likewise, an object pool is something we would always recommend you have in your project, via a plugin, on standby for when you start to spawn a lot of the same object. The dirty flag pattern is much more situational, though, and is only applicable when an object has lots of edits versus few read actions per second. Armed with these tools, you should be able to make a dent in the frame rate, destroying the spaghetti mess that all projects become before release. There are always more ways to optimize code beyond this too – some not quite so obvious – but the key is to remember that all data and all instructions are stored somewhere and actions using them require them to be moved, which takes time.

You can find the finished project with all the elements from this book completed on GitHub in the same place as the other chapters in the Complete branch. Feel free to create a fork from here and make your own improvements to each of these patterns:

https://github.com/PacktPublishing/Game-Development-Patterns-with-Unreal-Engine-5/tree/main/Complete

A final rule: Good code doesn't make a game good, but it does make your team better.

Index

www.packtpub.com

Subscribe to our online digital library for full access to over 7,000 books and videos, as well as industry leading tools to help you plan your personal development and advance your career. For more information, please visit our website.

Why subscribe?

- Spend less time learning and more time coding with practical eBooks and Videos from over 4,000 industry professionals
- Improve your learning with Skill Plans built especially for you
- Get a free eBook or video every month
- Fully searchable for easy access to vital information
- Copy and paste, print, and bookmark content

Did you know that Packt offers eBook versions of every book published, with PDF and ePub files available? You can upgrade to the eBook version at packtpub.com and as a print book customer, you are entitled to a discount on the eBook copy. Get in touch with us at customercare@packtpub.com for more details.

At www.packtpub.com, you can also read a collection of free technical articles, sign up for a range of free newsletters, and receive exclusive discounts and offers on Packt books and eBooks.

Other Books You May Enjoy

If you enjoyed this book, you may be interested in these other books by Packt:

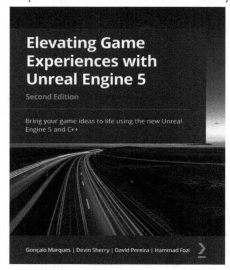

Elevating Game Experiences with Unreal Engine 5

Gonçalo Marques, Devin Sherry, David Pereira, Hammad Fozi

ISBN: 978-1-80323-986-6

- Create a fully functional third-person character and enemies
- Implement navigation with keyboard, mouse, and gamepad
- Program logic and game mechanics with collision and particle effects
- Explore AI for games with Blackboards and behavior trees
- Build character animations with animation blueprints and montages
- Polish your game with stunning visual and sound effects
- Explore the fundamentals of game UI using a heads-up display
- Discover how to implement multiplayer in your games

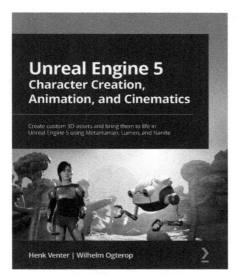

Unreal Engine 5 Character Creation, Animation, and Cinematics

Henk Venter, Wilhelm Ogterop

ISBN: 978-1-80181-244-3

- Create, customize, and use a MetaHuman in a cinematic scene in UE5
- Model and texture custom 3D assets for your movie using Blender and Quixel Mixer
- Use Nanite with Quixel Megascans assets to build 3D movie sets
- Rig and animate characters and 3D assets inside UE5 using Control Rig tools
- Combine your 3D assets in Sequencer, include the final effects, and render out a high-quality movie scene
- Light your 3D movie set using Lumen lighting in UE5

Packt is searching for authors like you

If you're interested in becoming an author for Packt, please visit authors.packtpub.com and apply today. We have worked with thousands of developers and tech professionals, just like you, to help them share their insight with the global tech community. You can make a general application, apply for a specific hot topic that we are recruiting an author for, or submit your own idea.

Hi!

We are Stuart Butler and Tom Oliver, authors of *Game Development Patterns with Unreal Engine 5*. We really hope you enjoyed reading this book and found it useful for increasing your productivity and efficiency.

It would really help us (and other potential readers!) if you could leave a review on Amazon sharing your thoughts on this book.

Go to the link below to leave your review:

<code>https://packt.link/r/1803243252</code>

Your review will help us to understand what's worked well in this book, and what could be improved upon for future editions, so it really is appreciated.

Best wishes,

Stuart Butler   Tom Oliver

Download a free PDF copy of this book

Thanks for purchasing this book!

Do you like to read on the go but are unable to carry your print books everywhere? Is your eBook purchase not compatible with the device of your choice?

Don't worry, now with every Packt book you get a DRM-free PDF version of that book at no cost.

Read anywhere, any place, on any device. Search, copy, and paste code from your favorite technical books directly into your application.

The perks don't stop there, you can get exclusive access to discounts, newsletters, and great free content in your inbox daily

Follow these simple steps to get the benefits:

1. Scan the QR code or visit the link below

https://packt.link/free-ebook/9781803243252

2. Submit your proof of purchase
3. That's it! We'll send your free PDF and other benefits to your email directly